Students' and Teachers' Values, Attitudes, Feelings and Beliefs in Mathematics Classrooms

Hanna Palmér • Jeppe Skott
Editors

Students' and Teachers' Values, Attitudes, Feelings and Beliefs in Mathematics Classrooms

Selected Papers from the 22nd MAVI Conference

Springer

Editors
Hanna Palmér
Mathematics Education
Linnaeus University
Växjö, Sweden

Jeppe Skott
Mathematics Education
Linnaeus University
Växjö, Sweden

ISBN 978-3-319-70243-8 ISBN 978-3-319-70244-5 (eBook)
https://doi.org/10.1007/978-3-319-70244-5

Library of Congress Control Number: 2017960172

Printed on acid-free paper

This Springer imprint is published by Springer Nature
The registered company is Springer International Publishing AG
The registered company address is: Gewerbestrasse 11, 6330 Cham, Switzerland

Preface

This book presents a selection of the papers presented at the 22nd international conference on affect in mathematics education, MAVI 22 (Mathematical Views). The conference was held at Linnaeus University, Växjö, Sweden, in September 2016.

The teaching and learning of mathematics is highly dependent on students' and teachers' values, attitudes, feelings, beliefs and motivations towards mathematics and mathematics education. The annual MAVI conference is an important venue for researchers interested in these affective issues, and at the conference in 2016, they were discussed by scholars from Sweden, Denmark, Finland, Germany, Canada, Switzerland and Israel.

The first MAVI conference was initiated by Erkki Pekhonen, University of Helsinki, and Günter Törner, University of Duisburg-Essen, and held in Germany in 1995. Ever since, it has been an important feature of the conference to provide opportunity for the participants to discuss their research for a fairly long period of time with colleagues in the field, as the same group of researchers work together for 2 or 3 days. This contributes significantly to take the field of affect in mathematics education forward and is an opportunity for researchers to make new national and international contacts.

While affect in mathematics education is a common interest for all participants in MAVI, the papers presented adopt a variety of theoretical perspectives, use a range of different methodologies and deal with different levels of education and mathematical contents. This variation is also represented by the papers selected for this volume. Between them, the papers discuss affect in mathematics in primary and secondary school as well as in programmes for teacher education and professional development; they deal with assessment issues, entrepreneurial competences and reasoning and proof; and they use frameworks that are relatively mainstream in the belief literature and others that adopt a more social stance. The selected papers have been peer reviewed and revised on the basis of the reviews and the feedback received during the conference.

The first three papers in this volume present studies of teachers in different kinds of professional development programmes. Liljedahl compares teachers' beliefs of proxies for learning (e.g. pretend to do a task) to their beliefs of what it means to teach and learn mathematics. The results indicate that teachers' views of mathematics (problem-solving view or toolbox view) influence if they see proxies for learning as their responsibility or as the responsibility of the students.

Palmér, Johansson and Karlsson report on changes in the teachers' role when entrepreneurial and mathematical competences are to be combined in teaching. These changes are explored in relation to how they seem to influence teachers' teaching of mathematics as well as students' possibilities to learn mathematics. The results show that increased emphasis on the entrepreneurial part of the combination may contribute to a more reform-oriented approach to teaching.

In the third paper, Rouleau and Liljedahl present a study where they deliberately introduce a tension in pre-service teachers' conception of timed drills and examine changes in their subsequent approach to teaching. Their findings suggest that the introduced tension provided the means for reflection on intent and resulted in a subsequent change in action.

In the next paper, Oksanen, Lahdenperä and Rämö present a study of Finnish university teaching assistants' professional identities. In the study, they use a questionnaire to categorise the metaphors with which the TAs describe themselves as professionals and find that most of the research participants use what the authors refer to as metaphors for being a *didactical expert*.

The succeeding papers focus on students in teacher education. Kihlblom Landtblom analyses Swedish prospective middle school teachers' conceptions of the mean, mode, and median. Based on a survey that asks how the teachers would explain these concepts to students, she finds that they generally rely on formal definitions and descriptions of procedures.

Larsen, Østergaard and Skott present a study of prospective teachers' understandings of and approaches to reasoning and proof (R&P). In a qualitative questionnaire, the research participants claim to find R&P important, but the results also suggest that they face considerable problems with these mathematical processes, almost irrespectively of their affective commitment to them.

Tsamir, Tirosh, Levenson and Barkai investigate secondary school prospective mathematics teachers' views of cases as a tool in teacher education. In general, the participants felt that the use of cases had an impact on their understanding of common mathematical errors and that cases based on mistakes they had themselves made during homework assignments were the most meaningful.

The next two papers mainly focus on the methods used to measure students' and prospective teachers' mathematics-related beliefs and self-efficacy. Sayers and Andrews report on the development and trial of an online survey instrument focused on uncovering prospective teachers' mathematics-related beliefs. The instrument was found to be reliable, and an exploratory factor analysis yielded seven interpretable belief dimensions. The interactions of these dimensions allude to groups of students likely to prove problematic during their programme.

Girnat takes his point of departure in the self-efficacy scale from PISA 2012. The scale was later used with Swiss 15-year-olds, and Girnat builds on the Swiss results to argue that (1) the scale is not unidimensional and (2) a more fine-grained analysis is needed that allows for different self-assessment in different mathematical subdomains.

Finally, four papers adopt different perspectives on students. In her paper, Roos discusses the experiences with assessment of Swedish lower secondary students, who are in need of some form of special assistance in mathematics, either because they perform much below or much above what is expected. In a small-scale qualitative study, she finds that current assessment practices significantly impact the students' experiences with school mathematics but that it does not support their mathematical learning.

Andrews and Nosrati compare the beliefs held by students in the Norwegian and Swedish secondary schools about a form of whole-class presentation performed by their teachers. One significant difference seems to be related to whether the students are in vocational or academic tracks.

In a second paper on Swedish and Norwegian secondary students, Nosrati and Andrews present the research participants' view(s) of a *typical mathematics lesson*. Based on a large interview study, they argue that there is much uniformity in how class time is spent and that in spite of curricular intentions to the contrary time use is highly structured.

Finally, Sumpter presents a study on secondary school students' mathematical reasoning when solving tasks about fractions. The results suggest that a focus on reasoning provides additional information about students' knowledge about fractions beyond standard error analysis.

We believe that the diversity of fields of interest and the multiplicity of theoretical and methodological approaches in these selected papers illustrate the innovative and inclusive spirit of the MAVI conference. Also, they reflect the constant development of research and knowledge within the community.

Växjö, Sweden

Hanna Palmér
Jeppe Skott

Contents

About the Authors

Paul Andrews is a professor of mathematics education at Stockholm University. He has extensive experience on research in comparative mathematics education, typically focusing on the didactics of school mathematics and critiques of international assessments like PISA. He is currently the director of both the Swedish Research Council-funded Foundational Number Sense (FoNS) project and a multinational examination of teacher education students' understanding of linear equations.

Ruthi Barkai is a researcher and teacher educator at the School of Education, Tel Aviv University, and a lecturer at Kibbutzim College. Her research interests include developing mathematical thinking among preschool students and their teachers, teachers' training, connections between proving and reasoning and professional development of pre-service and practising mathematics teachers at elementary and high school levels.

Boris Girnat holds a master's degree as a teacher for mathematics, philosophy and political science for lower and upper secondary level. He completed his doctorate in mathematics education with a study on teachers' beliefs concerning teaching geometry. Since 2011, he is a senior lecturer for mathematics education at the School of Teacher Education, Northwestern Switzerland. His areas of research are empirical studies on students' and teachers' beliefs, comparative studies and competence models.

Maria L. Johansson has a Ph.D. in mathematics and also one in mathematics education. Her Ph.D. in mathematics education focused on issues on mathematics in preschool, particularly with the professional development of preschool teachers and childcare workers. She is an associate professor at Luleå University where she works in teacher education. Her research interests are wide ranging, covering both mathematics and mathematics education for a range of ages.

Lena Karlsson is a teacher educator in preschool teacher education and primary school teacher education at Linnaeus University in Sweden. Her main interests are the development of mathematical thinking among students in mathematical difficulties, teacher training and connections between proving and reasoning as well as the inclusion of entrepreneurial competences in mathematics education.

Juulia Lahdenperä is a Ph.D. student in the Department of Mathematics and Statistics, University of Helsinki, Finland. She is conducting mixed methods research on how different course designs in undergraduate mathematics education affect students' experiences and achievement. The special focus of her research is in affective factors and the teaching-learning environment in relation to generating mathematical expertise.

Karin Kihlblom Landtblom is a lecturer in the Department of Mathematics and Science Education, Stockholm University. Her Ph.D. studies highlight the complexities in the teaching and learning of statistics, drawing on her previous experience as a teacher of mathematics at secondary school level and in more recent years with prospective teachers. Her research focus looks at how knowledge is constructed through tasks, real-life experience and language on averages, in years 4–6.

Dorte Moeskær Larsen is a Ph.D. student in mathematics education at the Southern University of Denmark. Her main research interests are teacher education, teacher development and mathematics teacher knowledge. Currently, she is working with the students' development of mathematical reasoning and how this can be assessed in primary schools.

Esther Levenson is a researcher and teacher in the Department of Mathematics, Science and Technology Education at Tel Aviv University and at the Kibbutzim College of Education. Her current research interests include fostering mathematical creativity among students and teachers, developing mathematical thinking among preschool children and their teachers, the connections between proving and reasoning and the role examples play in eliciting explanations.

Peter Liljedahl is a professor of mathematics education in the Faculty of Education and an associate member in the Department of Mathematics at Simon Fraser University in Vancouver, Canada. Dr. Liljedahl is the current president of the International Group for the Psychology of Mathematics Education and a senior editor for the *International Journal of Science and Mathematics Education*. Dr. Liljedahl is a former high school mathematics teacher who has kept his research interest and activities close to the classroom including creativity, insight and discovery in mathematics teaching and learning; the role of the affective domain on the teaching and learning of mathematics; the professional growth of mathematics teachers; mathematical problem-solving; numeracy; and engaging student thinking.

Mona Nosrati is a lecturer in mathematics education at the Norwegian Centre for Mathematics Education, Trondheim. Her teaching typically focuses on the professional development of serving teachers and the development of materials in support of it. Her research interests are broad and examine the development of mathematical understanding, particularly algebra and abstraction, neurological perspectives on the learning processes, and philosophical and sociological perspectives on various aspects of mathematics education.

Susanna Oksanen is a Ph.D. student in the Department of Teacher Education, University of Helsinki, Finland, and an in-service mathematics and German teacher teaching grades 7–9. Since 2012, she has focused on teachers' beliefs about mathematics and its teaching and learning. She has studied both in-service and preservice mathematics teachers and teaching assistants.

Hanna Palmér is associate professor in mathematics education at Linnaeus University, Sweden. Her main research interest is the professional identity development of mathematics teachers and the work with and learning of mathematics of children in preschool, preschool class and lower primary school. Of special interest is problem-solving in mathematics, digital technology in mathematics education and entrepreneurial competences.

Helena Roos is pursuing a Ph.D. in mathematics education at Linnaeus University, Sweden. Her current research addresses equity issues in terms of the meaning of and what influences inclusion in mathematics from a student perspective. Helena has been teaching in primary school and is currently a lecturer in mathematics education, teaching in both the teacher programme and special teacher programme which she also is responsible for.

Annette Rouleau is a Ph.D. student studying under Dr. Peter Liljedahl at Simon Fraser University in Vancouver, Canada. Her research interests are teacher education and professional development, with a focus on the role of tension in mathematics teachers' intentions, actions and professional growth.

Johanna Rämö is a university lecturer at the Department of Mathematics and Statistics, University of Helsinki, Finland. She has a Ph.D. in abstract algebra and is now focusing on university mathematics education development and research, especially by creating and studying new learning environments and fostering professional development of teaching assistants.

Judy Sayers is a senior lecturer in the Department of Mathematics and Science Education, Stockholm University. She is a former primary school teacher whose research and teaching typically address matters pertinent to the teaching and learning of mathematics by younger children. She is currently a member of the team, funded by the Swedish Research Council, researching the development of foundational number sense (FoNS) in England and Sweden.

Jeppe Skott is a professor of mathematics education at Linnaeus University, Sweden. His main research interests are social perspectives on teacher development. In recent years, the main focus has been on teachers' professional identity using a framework called patterns of participation (PoP). As part of developing PoP, theory networking has become another main concern. Currently, he is also working on mathematical reasoning in elementary teacher education.

Lovisa Sumpter is a reader (associate professor) and lecturer at the Department of Mathematics and Science Education, Stockholm University, Sweden. Her research interests are mathematical reasoning, affect and gender. Before doing her Ph.D., she was a teacher in mathematics and social science at upper secondary school level.

Dina Tirosh is a full professor of mathematics education in the Department of Mathematics, Science and Technology Education at Tel Aviv University. Her main areas of research are intuition and infinity, the theory of intuitive rules in mathematics and science, early childhood education and mathematics teacher education. Currently, she is a member of the editorial boards of several mathematics education journals and mathematics education series.

Pessia Tsamir is a full professor of mathematics education in the Department of Mathematics, Science and Technology Education at Tel Aviv University. Her main areas of research are intuition and infinity, the theory of intuitive rules in mathematics and science, the role of errors in mathematics learning and teaching, early childhood education and mathematics teacher education. She served as a vice president of PME.

Camilla Hellsten Østergaard is an associate professor at Metropolitan University College, Copenhagen. She teaches mathematics and mathematics education to prospective teachers. Her main research interest is teacher development during their pre-service education and especially lesson study. She is also working on statistical literacy and mathematical reasoning.

Relationship Between Proxies for Learning and Mathematically Related Beliefs

Peter Liljedahl

Abstract Students do not always behave the way we intend for them to. Sometimes they substitute the learning opportunities we offer them with behaviours that, on the surface, seem to be conducive to learning but in reality stand in place of learning. In this paper I explore teachers' beliefs of such behaviour, called proxies for learning, and compare them to teachers' beliefs of what it means to teach and learn mathematics. Results indicate that if teachers have a problem-solving view of mathematics, then they see such behaviours as their responsibility and theirs to fix. If teachers have a toolbox view of mathematics, they see these behaviours as inevitable and the responsibility of the students.

Keywords Beliefs • Belief cluster • Proxies for learning • Problem-solving

Proxies for Learning

In legal terms, a proxy is an "authority or power to act for another" (Merriam-Webster Dictionary, 2016) or "the authority to represent someone else" (Oxford Dictionary, 2016). More generally, a proxy is "something serving to replace or substitute for another thing" (Merriam-Webster Dictionary, 2016). In short, a proxy is something that *stands in place of* something else.

Within mathematics educational research, proxies exist implicitly within our use of metaphors, gestures, and manipulative. Even within the way we use data to stand in place of actual phenomena can be seen as a proxy.

> … proxy measurement is as old as measurement itself. (Scott, 1995)

More explicitly, the idea of proxies has emerged within my prior research both as a method and as a result. In my work on *Building Thinking Classrooms* (Liljedahl, 2016), I needed a way to measure student engagement. Having worked in the field of engagement before, I knew that there was no way to actually measure student

P. Liljedahl (✉)
Simon Fraser University, Burnaby, BC, Canada
e-mail: liljedahl@sfu.ca

H. Palmér, J. Skott (eds.), *Students' and Teachers' Values, Attitudes, Feelings and Beliefs in Mathematics Classrooms*,
https://doi.org/10.1007/978-3-319-70244-5_1

engagement. Within that specific context, the proxies for engagement that I chose were:

1. Time to task – how long it took to start
2. Time to first mathematical notation – how long it took to write something
3. Eagerness to start – to what degree students were eager to start
4. Discussion – to what degree there was discussion in the group
5. Participation – to what degree everyone in a group participated
6. Persistence – to what degree a group was willing to persist when challenged
7. Non-linearity of work – to what degree the work was chaotic
8. Knowledge mobility – to what degree do ideas move between groups

All of these are aspects of observable and measurable student behaviour that can stand in place of engagement as an indicator of student engagement.

More relevant to this paper, within our research on *studenting* (Liljedahl & Allan, 2013a, 2013b), we found that students' actions often subverted the intentions of the teacher. For example, in looking at student behaviour when they were asked to try to solve an example task on their own (Liljedahl & Allan, 2013a), some students either *stalled* (substituted doing the task for legitimate student behaviour like going to the bathroom) or *faked* (pretended to do the task). These students opted to replace an opportunity to learn with what we came to call studenting behaviours. More troubling were the students who, when asked to solve these example tasks as a way to check their understanding, *mimicked* rather than tried the task on their own. Mimicking is the mapping of the teacher's worked example into the task they are being asked to do. In doing so, these student replaced an opportunity to test their understanding with an activity that was not as conducive to learning. Taken together, these students' actions can be seen as *proxies for learning*. That is, these behaviours stand in place of legitimate learning behaviours.

These sorts of proxies for learning were also seen in our research on the ways students engaged in their homework (Liljedahl & Allan, 2013b), where student exhibited behaviours such as cheating, overreliance on help, and mimicking. Finally, in our unpublished research on the ways in which students take notes during class, the most prevalent behaviour was the mindless copying of notes. All of these behaviours are further examples of actions that stood in place of learning, or proxies for learning.

These examples of proxies for learning are not only in the purview of the students, however. Each of these student actions arose within the context of activities that the teacher asked the students to engage in. In some of these cases, such as when students were asked to solve an example task on their own, the teachers had goals that were conducive to learning, and it was the students who replaced these opportunities to learn with proxies. In other cases, such as the note taking, the teachers' goals were, in themselves, student activities that stood in the place of learning. Regardless, the teachers were, in all of these cases, implicated in the proxies for learning behaviours witnessed.

It is exactly this phenomenon that I am interested in. More specifically, I am interested in teachers' beliefs about their role in fostering, implicitly or explicitly, student proxies for learning behaviours. Further, I am interested in how these beliefs cluster with other teacher beliefs.

Belief Clusters

A persons' beliefs are not held in isolation from each other. Instead, they are made up of a loosely connected networks with links of varying degrees of strength and permanence (Nespor, 1987). These systems are not based on the substance of the beliefs but the psychological ways in which we hold them (Rokeach, 1968).

Green (1971) posits that there are three dimensions of belief systems:

- *Quasi-logical relationship:* Beliefs can be either primary or derivative. A derivative belief is a belief that is derived from a primary belief. Thompson (1992) offered the example of a teacher's belief that being able to clearly present mathematics may have a derivative belief that the teacher should then also be able to clearly answer questions asked of her.
- *Psychological strength:* Beliefs can be held either centrally or peripherally. Central beliefs are strong, often long-held, beliefs. Peripherally held beliefs are often new and unexamined beliefs. Green uses the metaphor of spatial order to talk about beliefs as being held with close or far to self.
- *Isolated clusters:* "Beliefs are held in clusters, as it were, more or less in isolation from other clusters and protected from any relationship with other sets of beliefs" (p. 48).

Chapman (2002) used this notion of belief systems (although she referred to them interchangeably as belief structures) as a framework for looking at teachers' changing practice. She concluded that we need to attend to teachers' central and primary beliefs if we are intending to influence teachers practice. With relation to belief clusters, Wilson and Cooney (2002) found through their deep study of a single teacher, Mr. Allen, that his understanding of mathematics and his beliefs about mathematics formed a belief cluster.

Meanwhile, Liljedahl, Rolka, and Rösken (2007) used a framework that differentiated mathematics into three philosophical beliefs to examine preservice elementary teachers' beliefs about mathematics as well as their beliefs about the teaching and learning of mathematics. These three beliefs can be summarized as a *toolbox*, *formalist*, and *problem-solving* (Dionne, 1984; Ernest, 1991; Törner & Grigutsch, 1994). In the *toolbox* view, mathematics is seen as a set of rules, formulae, skills, and procedures, while mathematical activity means calculating as well as using rules, procedures, and formulae. In the *formalist* view, mathematics is characterized by logic, rigorous proofs, exact definitions, and a precise mathematical language, and doing mathematics consists of accurate proofs as well as of the use of a precise and rigorous language. Finally, in the *problem-solving* view, mathematics is considered as a constructive process where relations between different notions and sentences play an important role. Here the mathematical activity involves creative steps, such as generating rules and formulae, thereby inventing or reinventing the mathematics. The results indicated that, both before and after the intervention, preservice teachers' beliefs about mathematics clustered with their beliefs about teaching and learning mathematics as an immersion in a problem-solving environment. Results also showed that these belief clusters shifted, either from a toolbox view or a formalist view, more towards a problem-solving view and as a result of the intervention.

In the research presented here, I am interested in examining the belief clusters of practicing mathematics teachers and seeing if their beliefs about mathematics cluster with their beliefs about teaching and learning mathematics. More importantly, I am also interested in how teachers view these proxies for learning and how those beliefs sit in relation to other beliefs.

Methodology

The research presented here is based on data collected from practicing secondary school mathematics teachers enrolled in a master's of mathematics education programme.

The Programme

The master's programme is designed specifically for practicing secondary school teachers. It is a cohort programme comprised of six courses, each taught over a 13-week period. As the enrolees are practicing teachers, classes run in the evening with one 4–5 h lesson per week. The course from which the data for this study emerges is the second course in the sequence.

The Course

The focus of this second course is to challenge teachers' notion of what it means to teach mathematics through immersion in a problem-solving context, experiences with an alternative teaching styles, exposure to literature on theories of learning, and exposure to research on the teaching of mathematics. As part of the course requirements, teachers were expected to experiment with specific teaching strategies within their own classrooms and reflect and report on these experiments within a journal. They were also expected to respond to a number of provocative prompts and questions within these same journals.

Of relevance, in the second week of the course, they were presented with the studenting research (Liljedahl & Allan, 2013a, 2013b) mentioned above.

The Participants

There were 16 teachers enrolled in the programme (and the course). Although teachers who complete a master's degree get a pay raise, my strong sense was that most of them had enrolled in the programme for their own edification and professional growth. The data for the research presented here comes from the journals of 12 of these teachers who ranged in teaching experience from less than 1 year to over 20 years, with an average of 7.4 years. Eight of the teachers taught only high school mathematics. The rest taught, in addition to mathematics, other school subjects from Science to Computer Science to Physical Education.

The Data

As mentioned, the teachers enrolled in the aforementioned course were expected to respond to a number of prompts in a journal. The particular prompts relevant to this paper are: *What is mathematics? What does it mean to learn mathematics? What does it mean to teach mathematics?* These prompts were asked in the first and last week of the course. Also in the last week of the course, they were asked to read over their journals and comment on what they notice to be the biggest changes in their thinking and their teaching and to comment on things that they will now always do in their teaching and things they will never again do in their teaching. In addition, in week two of the course, after seeing and discussing the aforementioned studenting research, they were asked to comment on proxies of learning they were witnessing in their students and in their teaching.

The journal entries pertaining to mathematics as well as the teaching and learning of mathematics were analysed using analytic induction (Patton, 2002) anchored in the three philosophical beliefs of mathematics: toolbox, formalist, and problem-solving. The journal entries pertaining to the proxies for learning were also analysed using analytic induction (Patton, 2002), this time anchored in the a priori codes from my studenting research (Liljedahl & Allan, 2013a, 2013b). Analytic induction, unlike grounded theory, begins with codes drawn from literature. Like grounded theory, however, analytic induction, through its constant comparative method, allows for the emergence of new codes and themes. And new themes did emerge out of the analysis of the journal entries pertaining to both beliefs about mathematics, teaching and learning mathematics, and proxies for learning.

Finally, a comparison was done to see if there existed any patterns between teachers' beliefs about mathematics and the teaching and learning of mathematics, on the one hand, and their beliefs of proxies of learning on the other.

Results and Analysis

In what follows I first present the analysis of the teachers' beliefs about mathematics and their beliefs about the teaching and learning of mathematics. I will then present the results of the analysis of the teachers' beliefs about proxies of learning. Finally, I will present an analysis of the relationship between these two sets of beliefs.

Beliefs About Mathematics and the Teaching and Learning of Mathematics

For the most part, the analysis of the journal entries using the aforementioned three philosophical beliefs of mathematics *flattened* the data. That is to say, it did not reveal much variance between the participants. Ten of the teachers viewed mathematics as problem-solving and two viewed it as a toolbox. Furthermore, the teachers expressed the same beliefs about mathematics at the end of the course as they did at the beginning of the course. This stands in stark contrast to the results of Liljedahl, et al. (2007) study of preservice teachers where there was a big shift towards a problem-solving view after immersion into a problem-solving environment. Given that most of the teachers came to the course already with a problem-solving view of mathematics coupled with the fact that the intervention was one of immersion in a problem-solving environment, this result is not surprising.

Having said that, a more thorough look at the way these ten teachers articulated their view of mathematics as problem-solving revealed a nuanced spectrum of ways to think about mathematics as a result of their experiences in the course. For example, both David and Nancy now view mathematics and mathematical problem-solving as an art.

> I have, in the last four years, viewed mathematics as solving problems. I've maybe viewed it as solving puzzles. But now I think of it more as an art. (David, week 13)

> … but my belief that mathematics is an art has been reaffirmed after this class. (Nancy, week 13)

Ellen now views mathematics as thinking.

> Mathematics is a way of thinking about patterns, shapes, numbers and concepts through discovery, invention, and problem solving. In order to learn mathematics, one needs to be engaged by personally thinking through a given problem. (Ellen, week 13)

Larry now views mathematics as the discovery of relationships through problem-solving.

> I still believe that the best way to teach is through guided problem solving where the teacher provides interesting material and hints along the way, but the students are responsible for the discovery and learning. (Larry, week 13)

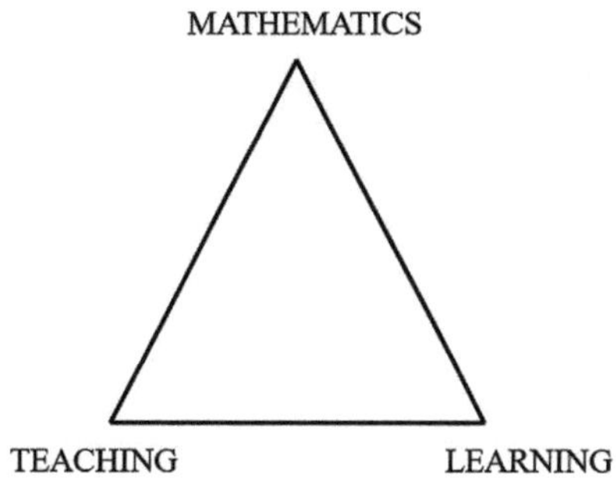

Fig. 1 Belief cluster of mathematics, teaching mathematics, and learning mathematics

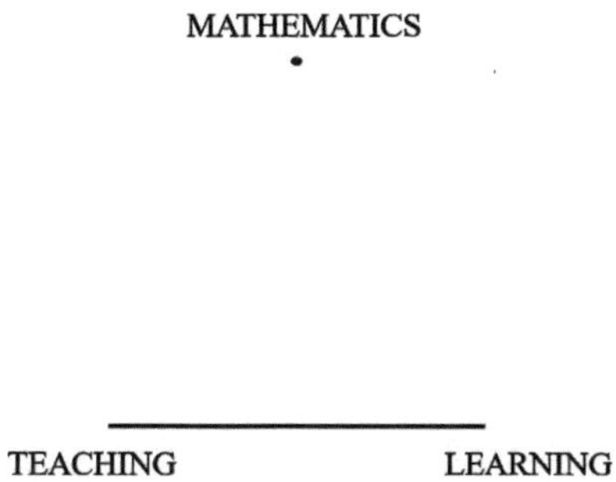

Fig. 2 Nancy's beliefs of mathematics, teaching mathematics, and learning mathematics

And, Barb now views mathematics as a set of tools for problem-solving with an emphasis on mathematics as problem-solving.

> Mathematics is the tools we develop through problem-solving. (Barb, week 13)

Looking now, specifically at the relationship between these teachers' view of mathematics, teaching mathematics, and learning mathematics, some interesting results emerge. First, ten of the teachers demonstrated very clear belief clusters encompassing these three beliefs (see Fig. 1), including the two teachers who had a toolbox view of mathematics.

> Mathematics is the study of developing mathematical thinking skills. Learning mathematics allows students to be able to connect mathematical ideas with new and existing ones. Teaching mathematics is to aid students to think mathematically by helping them conceptually understand mathematics. (Alison, week 13).

The exceptions to this were Nancy and Evan. While Nancy viewed mathematics as an art, she viewed learning mathematics as making connections between new knowledge and old knowledge, and to teach mathematics is the creation of environments that facilitate this (see Fig. 2). Based on this it can be said that for Nancy there is a belief cluster connecting her beliefs about the teaching of mathematics to her belief of the learning of mathematics, but not to her belief of mathematics. Evan, on the other hand, has belief cluster connecting mathematics and learning, but not teaching (see Fig. 3).

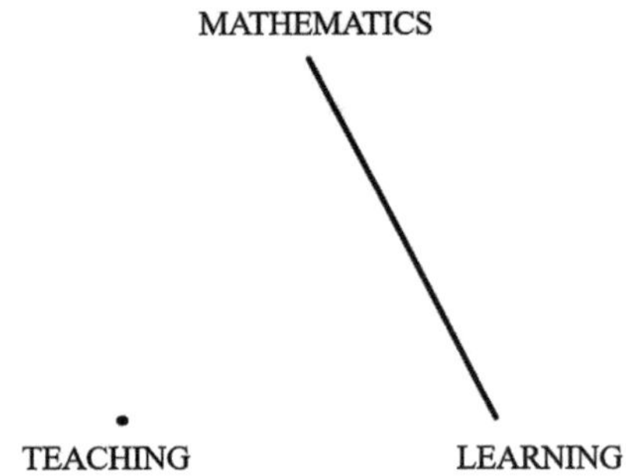

Fig. 3 Evan's beliefs of mathematics, teaching mathematics, and learning mathematics

Fig. 4 Typical flow of responsibility for students' proxy for learning behaviour

Proxies of Learning

With regard to the teachers' views of proxies of learning, some expected results emerged. Given the fact that the journal prompt came after the class in which we discussed the studenting research, I was expecting many of them to mention the obvious proxies for learning of faking, stalling, mimicking, and mindless note taking. And, as expected, each of the teachers mentioned this at some level. More interesting, however, were their comments as to wherein the responsibility for these behaviours lay. Nine of the teachers explicitly acknowledged that their students' proxy for learning behaviours were largely as a result of their teaching (see Fig. 4). Barb articulates this acknowledgment very nicely.

> Faking, stalling, mimicking and slacking are all evident on a daily basis. [..] I certainly have a role in enabling these proxies for learning; however I am starting to become more aware of them and taking some action to give my students some skills to move away from them. Certainly the 'you try one' style problems lend themselves to these actions and it makes me more aware of student actions. By spending the majority of time speaking to students I am not giving them time to think for themselves and construct their own relationships to the material. This teaches them that I will likely continue to offer the information or 'answers' and that they can wait for me to continue to do so. A learned behaviour of waiting for the result then occurs. By always offering answers or strategies students never truly take the ownership of the learning that needs to take place. We can then wonder whether or not they are truly learning at all. (Barb, week 2)

Their views on how to change this varied, but can be summarized either as a *start* or a *stop*. That is, two of the teachers felt that the way to curb this proxy behaviour was to start doing something different in their teaching—occupying their time so they can't stall or create more opportunities for collaboration so they cannot just follow the steps. Five of the teachers viewed the remedy as stopping doing things in their teaching. Barb, above, felt that she needed to stop speaking all the time as well as stop too quickly offering answers. There were also two teachers who had a combination of ideas of things to start doing and things to stop doing.

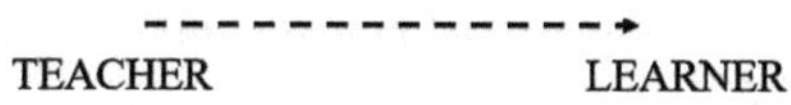

Fig. 5 Nancy's flow of responsibility for students' proxy for learning behaviour

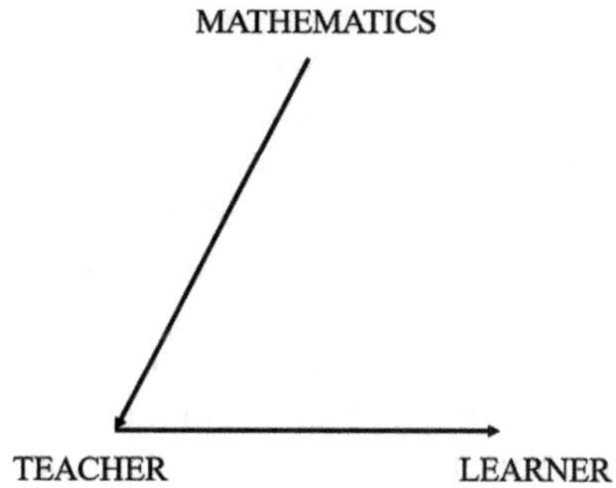

Fig. 6 Jessy and Abby's flow of responsibility for students' proxy for learning behaviour

Three teachers, however, did not see either starting or stopping something in their teaching as a possible solution. One of these is Nancy, who claimed that this proxy for learning behaviour was only partially due to her teaching (see Fig. 5).

> … there are numerous students who engage in mimicking behaviour, a proxy for learning. These students are good students, but I feel that their confidence in mathematics is lacking as they find it difficult to stray from the "notes" and create understanding on their own. If I changed the "you try" question to an application question these students struggled and then demonstrated "stalling" behaviour. I believe this is partially my fault as I always encourage them in my teaching to refer back to the notes when they are not sure what to do. (Nancy, week 2)

The two exceptions were for Jessy and Abby, both of whom saw their teaching as being responsible for the students' proxy behaviour. However, they also saw their teaching as being dictated by mathematics (see Fig. 6).

> Students tend to chat with others which usually results in a seat change. However, they keep trying to engage in other meaningless activities while taking notes including doodling and constantly checking time. I am enabling these proxies because my lesson is mostly teaching with notes on the board—students are sitting back listening to the lesson and taking notes down. (Abby, week 2)

Other than these two, no one else expressed feeling as though their teaching was an inevitable result of mathematics—an inevitable practice that they couldn't change despite recognizing that it was creating proxies for learning in their students.

Relating Beliefs and Proxies

When we look at the whole of the data together, an interesting relationship begins to emerge. There were only two teachers who did not have a problem-solving view of mathematics—Jessy and Abby. Although they had a belief cluster binding their view of mathematics to their beliefs of teaching and learning mathematics, that view was a toolbox view. That is, they viewed mathematics as a collection of things to be taught and learned. These two teachers were also the only ones who viewed the proxies of learning as an unavoidable consequence of the type of teaching that has to be done to teach their view of mathematics.

Nancy, although possessing a problem-solving view of mathematics, did not cluster this belief with her view of teaching and learning mathematics. At the same time, Nancy saw proxies of learning as only partially her responsibility as mimicking is an eventuality for her style of teaching.

The other teachers all had problem-solving belief cluster binding their beliefs of mathematics as well as the teaching and learning of mathematics and had a view that proxies of learning were within their domain of control and had found mechanisms to deal with this in their teaching.

The only outliers in all this was Evan, whose cluster did not include teaching of mathematics yet took full responsibility for the proxy behaviour of his students.

> I noticed some "proxies for learning". Of course, there were the usual suspects, such as students asking to be excused to use the washroom. But the most glaring proxy for learning was the method of learning itself. For example, in a Pre-Calculus 12 class this past week, my students gave very little input as I asked questions throughout the lesson. The reason seems obvious: they were waiting! They were waiting for me to write notes, so that they in turn could write down the exact same thing on their template notes. (Evan, week 2)

> I have really lost interest in having a class where I teach a certain curriculum topic by giving the students all the notes they will ever need, while all they do is simply copy down what I write. That kind of lesson at this point sounds very uninteresting and mindless for the students [..] I have also become wary of students doing seated work in class. (Evan, week 13)

Discussion

In the study presented here, proxies of learning were, for the most part, something that teachers viewed as being their responsibility in eliminating from their classroom. The only exceptions to this occurred for teachers who either did not have a problem-solving view of mathematics (Jessy and Abby) or did not have a belief about teaching and learning mathematics that clustered with their view of mathematics (Nancy). Put in the affirmative, all the teachers who saw the elimination of proxies of learning their responsibility also had a problem-solving view of mathematics that clustered with their beliefs about the learning of mathematics, and most also had the complimentary view of teaching mathematics.

Taken together, the results of the research presented here indicate that a problem-solving cluster of mathematics and the learning of mathematics, combined with a sense of responsibility for the elimination of proxies of learning, form a belief system. This result extends the work on belief clusters (Green, 1971) in general and mathematical belief clusters in particular (Liljedahl, et al., 2007). Moreover, it combines teachers' beliefs about mathematics and learning mathematics with explicit aspects of teaching practice. To my knowledge, this has never been done before.

References

Chapman, O. (2002). Belief structures and inservice high school mathematics teacher growth. In G. Leder, E. Pehkonen, & G. Törner (Eds.), *Beliefs: A Hidden Variable in Mathematics Education* (pp. 177–194). Dordrecht, Netherlands: Kluwer Academic Publishing.

Dionne, J. (1984). The perception of mathematics among elementary school teachers. In J. M. Moser (Ed.), *Proceedings of 6th Conference of the North American Chapter of the International Group for the Psychology of Mathematics Education* (pp. 223–228). Madison (WI): University of Wisconsin: PME-NA.

Ernest, P. (1991). *The philosophy of mathematics education*. Hampshire, UK: The Falmer Press.

Green, T. F. (1971). *The activities of teaching*. Tokyo: McGraw-Hill Kogakusha.

Liljedahl, P. (2016). Building thinking classrooms: Conditions for problem solving. In P. Felmer, J. Kilpatrick, & E. Pekhonen (Eds.), *Posing and solving mathematical problems: Advances and new perspectives*. New York, NY: Springer. https://doi.org/10.1007/978-3-319-28023-3_21

Liljedahl, P., & Allan, D. (2013a). Studenting: The case of "now you try one". In A. M. Lindmeier, & A. Heinze (Eds.), *Proceedings of the 37th Conference of the International Group for the Psychology of Mathematics Education* (Vol. 3, pp. 257–264). Kiel, Germany: PME.

Liljedahl, P., & Allan, D. (2013b). Studenting: The case of homework. In *Proceedings of the 35th Conference for Psychology of Mathematics Education – North American Chapter* (pp. 489–492). Chicago, USA.

Liljedahl, P., Rolka, K., & Rösken, B. (2007). Affecting affect: The re-education of preservice teachers' beliefs about mathematics and mathematics learning and teaching. In M. Strutchens & W. Martin (Eds.), *69th NCTM yearbook – The learning of mathematics* (pp. 319–330). Reston, VA: National Council of Teachers of Mathematics.

Merriam-Webster Dictionary. (2016). Retrieved from: http://www.merriam-webster.com/

Nespor, J. (1987). The role of beliefs in the practice of teaching. *Journal of Curriculum Studies, 19*(2), 317–328. https://doi.org/10.1080/0022027870190403

Oxford Dictionary. (2016). Retrieved from: http://www.oxforddictionaries.com/

Patton, M. Q. (2002). *Qualitative research and evaluation methods*. Thousand Oaks, CA: Sage.

Rokeach, M. (1968). *Beliefs, attitudes, and values*. San Francisco, CA: Jossey-Bass.

Scott, M. (1995). *Proxy measurement of adult basic skills: Lessons from Canada. Report Number NCAL-TR94-18 for the National Center on Adult Literacy*. Philadelphia, PA: National Center on Adult Literacy.

Thompson, A. (1992). Teachers' beliefs and conceptions: A synthesis of research. In D. A. Grouws (Ed.), *Handbook of Research on Mathematics Teaching and Learning* (pp. 127–146). New York: Macmillan.

Törner, G., & Grigutsch, S. (1994). Mathematische Weltbilder bei Studienanfängern – eine Erhebung. *Journal für Mathematikdidaktik, 15*(3/4), 211–252.

Wilson, S., & Cooney, T. (2002). Mathematics teacher change and development: The role of beliefs. In G. Leder, E. Pehkonen, & G. Törner (Eds.), *Beliefs: A Hidden Variable in Mathematics Education* (pp. 177–194). Dordrecht, Netherlands: Kluwer Academic Publishing.

Teaching for Entrepreneurial and Mathematical Competences: Teachers Stepping Out of Their Comfort Zone

Hanna Palmér, Maria Johansson, and Lena Karlsson

Abstract This paper reports on an educational design research study exploring the potential in combining the teaching of entrepreneurial and mathematical competences in Swedish primary schools. The focus in this paper, however, is not on the wholeness of this study but on changes in the teacher role when entrepreneurial and mathematical competences are to be combined in teaching – as expressed by the teachers themselves. Two of these expressed changes are "saying less" and "daring to let go of control". In the paper, these two changes are explored in relation to how they seem to influence these teachers' teaching of mathematics, and some implications are drawn regarding how their students' possibilities to learn mathematics may have been influenced.

Keywords Teaching • Entrepreneurial competences • Mathematical competences • Comfort zone • Educational design research • Primary school • Teacher change • Reform mathematics • Problem solving • Entrepreneurship

Introduction

Entrepreneurial and mathematical competences are two of the key competences the European Commission stresses as important in a society of lifelong learning. It is argued that entrepreneurial and mathematical competences will contribute to individuals' future success in society, no matter what kind of work they will do (EU, 2007). Mathematics is (most often) an unquestioned subject in school, and entrepreneurship, which is now being stressed by the European Commission, is also attracting greater interest in educational settings around the world. Entrepreneurship in this sense is not necessarily about starting companies, but

H. Palmér (✉) • L. Karlsson
Linnaeus University, Växjö, Sweden
e-mail: Hanna.palmer@lnu.se

M. Johansson
Luleå University of Technology, Luleå, Sweden

H. Palmér, J. Skott (eds.), *Students' and Teachers' Values, Attitudes, Feelings and Beliefs in Mathematics Classrooms*,
https://doi.org/10.1007/978-3-319-70244-5_2

rather it is viewed as an approach to education that gives children opportunities to develop abilities that characterize entrepreneurs. In Sweden, the government has a national strategy for entrepreneurship in the educational area (Regeringskansliet, 2009), and according to the Swedish primary school curriculum implemented in 2011, entrepreneurship is to permeate all teaching in primary school (National Agency for Education, 2011).

> The school should stimulate pupils' creativity, curiosity and self-confidence, as well as their desire to explore their own ideas and solve problems. Pupils should have the opportunity to take initiatives and responsibility, and develop their ability to work both independently and together with others. The school in doing this should contribute to pupils developing attitudes that promote entrepreneurship. (National Agency for Education, 2011, p. 11)

This paper reports on an educational design research study exploring the potential in combining the teaching of entrepreneurial and mathematical competences in Swedish primary schools. The basis for the study is that entrepreneurship is often taken for granted as something positive, but there are very few studies on entrepreneurial competences in subjects in general and in primary school in particular (Leffler & Svedberg, 2010). In the study, instead of taking an unconsidered stance, we try to investigate both possibilities and reservations regarding combining the teaching of entrepreneurial and mathematical competences. If entrepreneurship is to permeate all teaching in primary school, what does that imply for mathematics teaching and learning?

At the time of writing, the study had not yet been completed, so the results presented in this paper do not reflect the study as a whole. Instead, the explicit focus of this paper is on how primary school teachers involved in the study express changes in their teacher role now that they are required to combine entrepreneurial and mathematical competences in their teaching. Two of these changes have been expressed as "saying less" and "daring to let go of control". Thus, the changes discussed in the paper are changes expressed by the teachers, not changes identified by the researchers. However, in the paper these two expressed changes are explored in relation to how they seem to influence these teachers' teaching of mathematics, and some implications will also be drawn regarding how these changes may have influenced their students' possibilities to learn mathematics.

Entrepreneurial Competences

Entrepreneurial competences and entrepreneurship are stressed in both international and national educational contexts. The European Commission argues that entrepreneurship is important in a society of lifelong learning. Entrepreneurial competences are not only for older students but rather should be developed through continuous learning adapted to students' age (EU, 2007). One question to ask is which competences are considered to be entrepreneurial? Based on an overview, Holmgren and

From (2005) emphasize competences such as creativity, innovation, risk taking, opportunity spotting, self-motivation and ability to cope with uncertainty as entrepreneurial competences, which is similar to the definitions stated by the European Community.

Based on the European Community, on the Swedish national curriculum and on research literature on entrepreneurship (Jeffrey & Craft, 2004; Holmgren & From, 2005; Leffler & Svedberg, 2010; Sarasvathy & Venkataraman, 2011), the study presented in this paper has focused on the following six entrepreneurial competences: creativity, ability to take responsibility, ability to take initiative, tolerance for ambiguity, courage and ability to collaborate, and they are defined as follows:

Creativity is about finding new, for the individual, solutions to new and old problems and is well researched as positive in both entrepreneurial and mathematics education research. The *ability to take responsibility* regards both oneself and others. To be able to take responsibility, students must be given autonomy, which is about passing back control to the learner. The *ability to take initiative* is about being proactive, which, just like the ability to take responsibility, is dependent on autonomy. *Tolerance for ambiguity* is about acting in situations where a task is not fully understood or where the scope for action is not fixed in advance. *Courage* is about stepping into situations in which the individual is not fully comfortable; thus it is about leaving the comfort zone. Finally, the *ability to collaborate* involves both sharing and absorbing ideas and knowledge.

Mathematical Competences

Mathematical competences are also stressed as important in a society of lifelong learning, especially with a focus on problem-solving in everyday situations (EU, 2007). In the Swedish national curriculum, mathematics is described as a "creative, reflective, problem-solving activity" (National Agency for Education, 2011, p. 62). Describing mathematics in terms of creativity, reflection and problem-solving can be seen in contrast to a national inspection of mathematics teaching undertaken in 2009, which showed that mathematics teaching was dominated by individual calculating, offering limited opportunities for students to develop their ability to solve problems (Swedish Schools Inspectorate, 2009). On the basis of those documents, problem-solving in mathematics has been especially – but not solely – emphasized in the study. In line with research (Cai, 2010; Lesh & Zawojewski, 2007), problem-solving is described in the national curriculum both as a purpose (an ability to formulate and solve problems) and a strategy (a way to acquire mathematical knowledge). In the study, both of these have been focused on; students have worked with problem-solving tasks they did not know in advance how to solve, and they therefore had to develop new (for them) strategies, methods and/or models to solve the tasks. In such an approach, students have to investigate new ways of thinking where creativity and tolerance for ambiguity are often emphasized as important abilities (Lesh & Zawojewski, 2007). Thus, a problem-solving approach in mathematics teaching has similarities with education *for* entrepreneurship as presented above.

Theoretical Perspective

In combining two different subjects – or rather two different sets of competences – and analysing what effect this combination has on the students' possibilities to learn (specifically the mathematical competences in this case), we need to analyse how the arrangement of the different mathematical activities change when the entrepreneurial competences are added in the lessons. According to Rogoff (2003) and Wertsch (1998), individuals' (e.g. teachers' and students') possibilities for learning depend on the activities they participate in, where learning implies changes in how they participate in these activities. Thus, learning is about becoming by participating in practices. This means that we have taken on a sociocultural perspective on learning. The changes expressed by primary school teachers regarding their teacher role now that they are required to combine entrepreneurial and mathematical competences in their teaching can be understood as their learning by participating in, what are for them, new activities. Similarly, these activities change what the students are invited to participate in, thereby also influencing their possibilities to learn. As mentioned, the focus of this paper is on the changes expressed by the primary school teachers regarding their teacher role now that they are required to combine entrepreneurial and mathematical competences in their teaching, with only some implications for the teachers' teaching of mathematics and students' possibilities to learn mathematics.

The Study

Nine researchers from mathematics education and entrepreneurship as well as approximately 30 teachers from eight primary schools were involved in the study, which was conducted through educational design research. Primary school in Sweden implies students from 6 to 12 years of age. This paper will focus on the teachers at four of these schools, the schools where the authors of this paper were the participating researchers. The 21 participating teachers at these schools were educated as generalists, teaching several subjects, one of which is mathematics. Before becoming involved in the study, the teachers from these four schools had participated in a national professional development programme named *Boost for Mathematics*. The programme was initiated by the government in 2012 with the aim of improving mathematics teaching and thereby students' learning. The programme has been developed by researchers and is organized around teacher collaboration, where teachers work in groups with experienced tutors. Within this programme the teachers at these schools had focused especially on problem solving.

The teachers were interviewed before and after the design research study. The interviews were open-ended, based on a template. All interviews were recorded. Design research is not a fixed method but a genre of inquiry within which solutions to practical and complex educational "problems" are developed through an iterative process of designing, implementing and evaluating lessons (McKenney & Reeves, 2012). The focus of the designed lessons was on combining the teaching of entrepreneurial and mathematical competences. Often, design research aims at finding solutions to educational problems and figuring out what teaching should look like to reach a desirable situation. This study, however, was more explorative since we did not know whether combining the teaching of entrepreneurial and mathematical competences was desirable or not.

The five mathematical competences emphasized in the national curriculum – with special emphasis on problem-solving – together with the six entrepreneurial competences presented earlier framed the design of the lessons. Each iterative design cycle included preparations for a mathematics lesson, implementation of this lesson and finally a retrospective analysis of the lesson. The researchers participated actively in the preparation and evaluation of the lessons and passively during the lessons taking notes. The joint evaluation was made using an evaluation form that focused on both entrepreneurial and mathematical competences as well as on possible connections between them. The focus of the evaluation was in line with the sociocultural perspective on what had been possible for the students to learn during the lesson and how the combination of entrepreneurial and mathematical competences may have contributed, positively or negatively. After the evaluation, the next lesson was planned, and the iterative process continued in this manner throughout one school year.

The result presented in this paper is based on empirical material from the follow-up interviews with the teachers together with observations and evaluations of two lessons. Two of the questions in the follow-up interview were "What characterizes a project lesson for you?" and "Has the project influenced your mathematics teaching, and if it has, how?" The analysis of the interviews was done using grounded theory methods, which implies building and connecting categories grounded in the empirical material by using codes (Charmaz, 2006). This way of coding does not involve the use of pre-constructed codes, but instead entails labelling the empirical material, line by line, with as many codes as possible (Kelle, 2007). When coding the follow-up interviews from one of the schools, two of the categories developed were "saying less" and "daring to let go of control". Segments of similar kinds were then looked for in the interviews from the other three schools.

Then, lessons where the teachers, during or after, had talked about "saying less" and/or "daring to let go of control" were analysed as empirical examples illustrating what the teachers meant with these expressions. These lessons were also analysed based on students' opportunities to learn mathematics. Since the students were not tested regarding their mathematical knowledge before and after the intervention, we cannot analyse their learning but only opportunities for learning. The lessons were analysed based on the students' opportunities to learn the five mathematical competences emphazised in the national curriculum (formulating and solving problems; using and analysing mathematical concepts; choosing and using appropriate methods to perform calculations and solve routine tasks; applying and following mathematical reasoning; and using mathematical forms of expression to discuss, reason and give an account of questions, calculations and conclusions). (Analysis was also done with respect to the entrepreneurial competences, but those are not focused on in this paper).

Results

Extracts from the interviews will first be presented, followed by an analysis of example lessons. Extracts and examples are not to be seen as the entire basis of the analysis but as examples of empirical instances labelled as "saying less" or "daring to let go of control".

Saying Less

Several of the teachers talked about how the project made them, or forced them to, give less instruction to the students when introducing the lessons. They expressed this as "saying less" and handing over the acting space to the students. To make it possible for the students to take initiative and be creative, and to challenge students' tolerance for ambiguity, the teachers used more open questions and gave less instruction on how to solve tasks.

Teacher, grade 4:	More open questions. I let them discover more by themselves. [...] I have a clear mathematical idea. But don't give them too much because then the challenge disappears.
Teacher, grade 3–4:	My approach towards students and learning has changed. I have higher expectations but try not to steer too much, work more with open questions; you don't need to give them all the instructions at once.
Teacher, grade 6:	Open for several different ways to reach the goal.

Example of a Lesson Illustrating "Saying Less"

In a grade 4 class, the students are working on a project on schoolyards. As part of this project, the teacher has an idea of the students comparing the sizes of different schoolyards in the municipality. She wants the students to explore how to calculate area for figures with irregular sides. The students are divided into groups of three, and each group is handed a map of a schoolyard. In the previous lesson, the students had worked with scale, and when introducing the schoolyard task, the teacher starts to repeat yesterday's talk about scale. Then she suddenly stops, turns towards the researcher and says, "Now I'm saying too much, aren't I?" She then writes "1 cm = 1 m" on the board and says to the students, "Now calculate and compare the size of the schoolyards". Thus, she does not say anything about how to calculate area for figures with irregular sides nor about how to compare the sizes. Different groups calculate the area in different ways. Some groups split their schoolyard into small pieces where each piece is a figure for which they know how to calculate the area. Some groups draw a large rectangle approximately around their schoolyard and then colour pieces of the schoolyard left on the outside of the rectangle. They then do the same with pieces that are inside the rectangle but are not part of the schoolyard. Finally, they compare to see if the coloured areas cancel each other out. When all groups have calculated the area of their schoolyard, they are to compare their results. The teacher makes a table on the board based on students' results. Each group explains how they solved the task, and the teacher highlights similarities and differences in their strategies. The different schoolyards vary greatly in size, and at first the students say this is unfair. Then one student says that "to be able to know if it is fair you must know how many students there are in each school". As homework, the groups are to find out the number of students enrolled in the school they are working with. The following day the whole-class discussions continue, and the table is extended to include a column headed "students" and another column headed "square metres/student".

Analysis Example of Lesson "Saying Less"

An evaluation of the lesson in relation to students' opportunities for learning the five mathematical competences emphasized in the national curriculum indicates possibilities for students to develop their ability to solve a problem using mathematics and also to assess selected strategies and methods (how to figure out the area, whole-class discussion of strategies); use and analyse mathematical concepts and their interrelationships (peripheral, scale, area, square metre, square metres/student); choose and use appropriate mathematical methods (calculate area, scale, square metre); apply and follow mathematical reasoning (whole-class discussions based on students' ideas and calculations); and use mathematical forms of expression to discuss, reason and give an account of questions, calculations and conclusions (producing, presenting and evaluating calculations and tables). To summarize, the lesson where the teacher tries to adjust in line with the entrepreneurial competences by "saying less" becomes a lesson where the students have opportunities to learn all the mathematical competences emphasized in the national curriculum.

Daring to Let Go of Control

Several of the teachers also talked about how the project made them, or forced them, to let go of control in the classroom. To make it possible for the students to take their own initiatives, be creative and take responsibility, the teachers were forced to give the students increased influence. At the end of the project, they talked about this as something positive.

Preschool class:	It is about us daring to try. To give the students more influence. Dare to let go of the control and try.
Grade 1:	I dare to trust the students more. Release them. […] I don't prepare in as much detail as before. […] The students have more influence. […] Before, I always needed to have all the answers, now we find them together.
Grade 2:	What is true for the students also applies to me. I have become more open to unplanned change. I dare to take in students' suggestions, I dare to go out in the quagmire and see where it carries us.
Grade 5–6:	I dare more; letting go of control is important! It doesn't matter if you fail, both we and our students learn from that.

Example of a Lesson "Daring to Let Go of Control"

At one of the schools, all involved teachers from grades 1 to 6 choose to work with what they refer to as Fermi problems to promote tolerance for ambiguity. These tasks are open problems where exact answers are difficult or impossible to arrive at, so estimates must be made instead, based mainly on known facts or facts that can be easily found (Flognman, 2011). This lesson is also an example of "saying less" as the teachers in their joint preparation talk about giving the students very few instructions. The teachers themselves say that their students usually get very clear instructions and they want to see how the students handle the open problems together with few instructions. The teacher in grade 2 starts her lesson by showing the students a large bottle with small candies inside. The questions posed to the students are, "How many candies are there in the bottle? We are not allowed to open the bottle. How can we solve this?" The teacher splits the class into groups with three students in each. After a while, each group comes up with a suggestion for how to solve the task. One group suggests that they can go to the local shop and buy a bag of the same kind of candy as in the bottle, count the pieces of candy in the bag and then figure out the number of candies in the bottle by comparing with the number in the bag. The teacher gives them money, and the group goes to the local shop and buys a bag of the same kind of candy as in the bottle. Back in the classroom, the students weigh the candy even though the weight was written on the bag. They want to be sure. Then they count the pieces of candy. After that, they ask if the teacher has an empty bottle similar to the one she has shown with candy inside, and she tells them to look in the teachers' office. The students find an empty bottle of similar size and they put the bought candy into this bottle, and then they estimate how many more bags of candy they would need to fill the bottle. After the lesson, the teacher says that she would never have let the students continue with their own initiative – to go to the local shop – before becoming involved in the projects on combining entrepreneurial and mathematical competences.

Analysis Example of Lesson "Daring to Let Go of Control"

An evaluation of the lesson in relation to students' opportunities for learning the five mathematical competences emphasized in the national curriculum indicates possibilities for students to develop their ability to solve a problem using mathematics and also to assess selected strategies and methods (quantity of candy, discussion of strategies); use and analyse mathematical concepts and their interrelationships (estimation, weight versus quantity, volume versus quantity); choose and use appropriate mathematical methods (count, weigh); apply and follow mathematical reasoning (whole-class discussions based on students' ideas and calculations); and use mathematical forms of expression to discuss, reason and give an account of questions, calculations and conclusions (producing, presenting and evaluating calculations of the amount of candy). To summarize, the lesson where the teacher tries to adjust in line with the entrepreneurial competences by "daring to let go of control" becomes a lesson where the students have opportunities to learn all the mathematical competences emphasized in the national curriculum.

Discussion

Several national and international studies report how curricular reforms in mathematics education seldom result in changes in the teaching of mathematics in schools. This is often explained with reference to teachers having difficulties with identifying the meaning of the messages of the different reforms (Boesen et al., 2014; Ross, McDougall & Hogaboam-Gray, 2002). The study presented in this paper does not focus on curricular reforms in mathematics education but on possibilities and reservations regarding combining the teaching of entrepreneurial and mathematical competences. The explicit focus of this paper has been on how primary school teachers involved in the research project express changes in their teacher role when entrepreneurial and mathematical competences are to be combined in their teaching.

Two of these expressed changes are "daring to let go of control" and "saying less". These two changes entail the teachers stepping out of their comfort zone. This could be seen as the teachers changing the context of the mathematical activities and hence also changing the possibilities for the students to learn the mathematical competences. (As mentioned, based on the sociocultural perspective on learning in this study, students' possibilities for learning depend on the activities we invite them to participate in.) They do this without changing how they talk about the mathematical context; rather, they add the entrepreneurial competences to the activities as seen in the analysed examples.

In the paper, these two expressed changes have been explored in relation to how they seem to influence the teaching of mathematics. It seems like "daring to let go of control" and "saying less", which the teachers expressed as changes to fulfil the entrepreneurial part of the combination, actually direct their mathematics teaching in a reformative direction. It can be argued that the designs of the lessons described above have been known and promoted in mathematics education for a long time. However, it was not until creativity, tolerance for ambiguity, courage, ability to take initiative, ability to collaborate and ability to take responsibility were introduced as important competences in themselves that these teachers planned these kinds of mathematics lessons. As mentioned, the teachers from these four schools had participated in the national professional development programme named *Boost for Mathematics*, and within this programme they focused especially on problem-solving. However, the kind of mathematics teaching that is exemplified in this paper is, according to the teachers, not a result of thinking about mathematics education but a result of trying to include entrepreneurial competences in the mathematics lessons. It seems that a focus on entrepreneurial competences in mathematics lessons has developed the lessons towards the creative, reflective, problem-solving activities described in the Swedish curriculum (National Agency for Education, 2011). Thus, changes that had been absent before and explained as teachers having difficulties identifying the meaning of the messages of different reforms are now being realized, but without the teachers having mathematics teaching in mind.

References

Boesen, J., Helenius, O., Bergqvist, E., Bergqvist, T., Lithner, J., Palm, T., et al. (2014). Developing mathematical competence: From the intended to the enacted curriculum. *The Journal of Mathematics Behavior, 33*, 72–87. https://doi.org/10.1016/j.jmathb.2013.10.001

Cai, J. (2010). Commentary on problem solving heuristics, affect, and discrete mathematics: A representational discussion. In B. Sriraman & L. English (Eds.), *Theories of mathematics education: Seeking new frontiers* (pp. 251–258). London: Springer. https://doi.org/10.1007/978-3-642-00742-2_25

Charmaz, K. (2006). *Constructing grounded theory. A practical guide through qualitative analysis*. London: SAGE Publications Ltd. https://doi.org/10.7748/nr.13.4.84.s4

EU. (2007). *Nyckelkompetenser för livslångt lärande. En europeisk Referensram*. Luxembourg, Europe: Europeiska gemenskaperna.

Flognman, C. (2011). *Fermiproblem och klassrumskultur. Nämnaren Nr3*. Göteborg, Sweden: NCM.

Holmgren, C., & From, J. (2005). Taylorism of the mind: Entrepreneurship education from a perspective of educational research. *European Educational Research Journal., 4*(4), 382–390. https://doi.org/10.2304/eerj.2005.4.4.4

Jeffrey, B., & Craft, A. (2004). Teaching creatively and teaching for creativity: Distinctions and relationships. *Educational Studies, 30*(1), 77–87. https://doi.org/10.1080/0305569032000159750

Kelle, U. (2007). The development of categories: Different approaches in grounded theory. In A. Bryant & K. Charmaz (Eds.), *The SAGE handbook of grounded theory* (pp. 191–213). London: SAGE Publications. https://doi.org/10.4135/9781848607941.n9

Leffler, E., & Svedberg, G. (2010). *Skapa och våga. Om entreprenörskap i skolan*. Stockholm: Skolverket.

Lesh, R., & Zawojewski, J. (2007). Problem solving and modeling. In F. K. Lester (Ed.), *Second handbook of research on mathematics teaching and learning* (pp. 763–799). Charlotte, NC: National Council of Teachers of Mathematics & Information Age Publishing.

McKenney, S., & Reeves, T. (2012). *Conducting educational design research*. London: Routledge. https://doi.org/10.4324/9780203818183

National Agency for Education. (2011). *Curriculum for the primary school, preschool class and leisure time center 2011*. Stockholm: National Agency for Education.

Regeringskansliet. (2009). *Strategi för entreprenörskap inom utbildningsområdet*. Stockholm: Regeringskansliet.

Rogoff, B. (2003). *The cultural nature of human development*. Oxford, England: Oxford University Press. https://doi.org/10.1037/e612832007-002

Ross, J. A., McDougall, D., & Hogaboam-Gray, A. (2002). Research on reform in mathematics education, 1993–2000. *Alberta Journal of Educational Research, 48*(2), 122–138.

Sarasvathy, S. D., & Venkataraman, S. (2011). Entrepreneurship as method: Open questions for an entrepreneurial future. *Entrepreneurship: Theory & Practice, 34*, 113–135. https://doi.org/10.1111/j.1540-6520.2010.00425.x

Swedish Schools Inspectorate. (2009). *Undervisningen i matematik—utbildningens innehåll och ändamålsenlighet. Kvalitetsgranskning rapport 2009:5*. Stockholm: Skolinspektionen.

Wertsch, J. V. (1998). *Mind as action*. New York: Oxford University Press. https://doi.org/10.1093/acprof:oso/9780195117530.001.0001

Creating Tension Between Action and Intent

Annette Rouleau and Peter Liljedahl

Abstract Pre-service teachers come to mathematics method courses with well-established conceptions of what it means to teach and learn mathematics. Images of teaching reinforced by their own lived experiences shape their pedagogy. This can be problematic for a teacher educator for whom it may be necessary to offer a way of reframing traditional notions of teaching and learning. The research presented here examines that process of reframing. In this study we deliberately introduce a tension in pre-service teachers' conception of timed drills and examine the resulting process of transition they undergo. Our findings suggest that the introduced tension provided the means for reflection on intent and resulted in a subsequent change in action.

Keywords Tension • Pre-service teachers • Teacher education • Timed drills • Multiplication • Folkways

Motivation for the Study

> Math facts are a very small part of mathematics but unfortunately students who don't memorize math facts well often come to believe that they can never be successful with math and turn away from the subject… For about one third of students the onset of timed testing is the beginning of math anxiety. (Boaler, 2015).

First Author Narrative

It was my first foray into teaching an elementary mathematics method course for pre-service teachers. Wanting to gauge their thoughts around the teaching and learning of basic facts, I broached the subject of timed drills. With timed drills being a

A. Rouleau (✉) • P. Liljedahl
Simon Fraser University, Burnaby, BC, Canada
e-mail: annetterouleau@gmail.com

H. Palmér, J. Skott (eds.), *Students' and Teachers' Values, Attitudes, Feelings and Beliefs in Mathematics Classrooms*,
https://doi.org/10.1007/978-3-319-70244-5_3

common practice among elementary teachers, it was not surprising that a survey of my pre-service teachers revealed that they had all experienced timed drills as students and that the majority intended to utilize them in practice. While not unexpected, this was problematic for me. With their emphasis on memorization and speed, timed drills are at best ineffective and, at their worst, potentially harmful to students.

Mindful that 'telling' my pre-service teachers about the detrimental effects of timed drills would have little impact on their practice, I decided to replicate a learning experience designed to reframe conceptions of timed drills (Liljedahl, 2014). Gathering my pre-service teachers around me, I told them we were going to have a multiplication drill—they would be required to respond verbally to a random multiplication question within an allotted time frame. Correctly answering the question would allow them to 'sit out', essentially completing their role in the intervention; incorrect or slow answers meant the pre-service teacher had to continue playing. By default, the last person standing would be the 'loser'. Although a handful appeared excited by the prospect of competing, the majority of the pre-service teachers were visibly anxious. The overwhelming relief in the room when I announced that this was a ploy and they would not actually have to answer was palpable. And the ensuing discussion was rich and reflective—they underwent a transformative experience that they felt compelled to share.

The comments during the debrief reflected a newfound awareness that the fear and anxiety experienced by the group would likely be the same emotions that the majority of children in their classroom would feel in a similar situation. This was expressed by both those who feared the intervention and by those who were excited by it. Their later journal entries reaffirmed that theme and also revealed that they would no longer consider using timed drills in their classrooms. This was despite the majority previously indicating that they had 'no issue' with timed drills.

This shift from planning to use timed drills to an avowal never to use them was intriguing. It seemed clear that the intervention had achieved its intended goal of raising an awareness in the pre-service teachers that caused them to reflect and reconsider implementing timed drills in their future classrooms. What was less clear, however, was the mechanism by which this transition occurred. The aim of this paper is to explicate that transition process. Thus, our research question is: What is the process through which pre-service teachers shift from acceptance of an established teaching practice to a determination never to use it?

Theoretical Background

Mathematics for many people is commonly associated with being able to get the correct answer quickly without the need for conceptual understanding (Boaler, 2015). It is not surprising then that timed drills, in which students are required to answer basic fact questions in timed tests, are an accepted practice among elementary teachers (Kling & Bay-Williams, 2014). Drills are completed individually or, more commonly, as a group activity where the students are required to answer in front of their peers. With their emphasis on memorization and speed, these drills are unnecessary and damaging (Boaler, 2015; Harper & Daane, 1998). Instead, research suggests that effective teaching practices are those that promote conceptual understanding (NCTM, 2014; Tirosh & Graeber, 2003).

Yet the use of timed drills persists, and a visitor to any Canadian elementary classroom is likely to encounter timed tests. Timed drills are iconic of established teaching practices, which have come to be commonly accepted, and little conscious thought is put into their continued use and implementation (Buchmann, 1987). Referred to as 'folkways' of teaching, they cut across the experiences of students and are built up through collective participation. Accepted as the way mathematics is taught and learned, folkways are 'capable of being practiced without understanding their point or efficacy, the folkways are widespread and emblematic, expressing in symbol and action what teaching is about' (Buchmann, p. 7). In order to disrupt the universal acceptance of folkways such as timed drills, it is necessary to provide a means of raising awareness and reflection.

One way to achieve this is to introduce a tension. Typically a byproduct of teaching, tensions are described as the inner turmoil experienced by teachers. They are the unintended yet inevitable consequence for teachers who find themselves pulled in differing directions by competing pedagogical demands. However tensions can be useful for those who accept the conflicts and use them to shape identity and practice (Lampert, 1985). It is tension that often propels teachers towards professional development and provides the impetus to improve their practice (Rouleau & Liljedahl, 2015).

However, teachers are not always attuned to these tensions, and subsequently there is no reflection to provide that impetus (Berlak & Berlak, 1981). As with the pre-service teachers, whose previous experience with the established folkway of timed drills had deflected any awareness of a tension, a tension may need to be deliberately introduced for a change in practice to occur (Liljedahl, 2014). Berlak and Berlak (1981) suggest that because a person is capable of being made aware of tensions, they are capable of altering their practice. It is important to keep these two notions distinct; a teacher educator can provide the opportunity for change, but the agency of change must lie with the pre-service teacher. 'Effective change is something that people do to themselves; more radically, but more aptly when investigated closely, change is something that happens to people who adopt an enquiring stance towards their experience' (Mason, 2002, p. 143).

A framework for both identifying and understanding tensions emerged from the work of Berry (2007). Isolating six pairs of interconnected tensions, Berry used these as a lens to examine her practice. These pairs of tensions are (1) telling and growth, (2) confidence and uncertainty, (3) safety and challenge, (4) valuing and reconstructing experience, (5) planning and being responsive and (6) intent and action. The last of these—the tension 'between working towards a particular ideal and jeopardising that ideal by the approach chosen to attain it' (p. 32)—is the tension that is most relevant to the phenomenon introduced at the beginning of this article. Until participating in the timed drill intervention, the pre-service teachers had expressed no conflict between their *intent* of having students learn basic facts through their chosen *action* of timed drills. It was difficult for them to recognize the pitfalls inherent in habitual ways of practice even though these ways were actually working against their intended goals for their students' learning. The implementation of the intervention introduced the tension, which resulted in them making the transition from wanting to utilize timed drills to never wanting to use them.

In what follows we use Berry's (2007) tension pairing of intent and action to frame the transition process experienced by the pre-service teachers. In doing so, we examine the process of moving from their acceptance of an established practice to a determination that they will never use it in their practice.

Method

The participants for this study were 69 pre-service teachers enrolled in two sections of a fourth year elementary mathematics method course taught by the first author. As the lead-up to a lesson on the teaching and learning of basic facts, the pre-service teachers experienced an intervention designed to cause tension between their professed intent of having their students learn basic facts and the action of incorporating timed drills into their teaching practice. A whole group debrief of the experience followed immediately. The pre-service teachers were then required to complete a reflective journal entry composed of prompts modelled on Gibbs' reflective cycle (1988). First, they were asked to describe the experience, then describe what they were thinking and feeling, and finally, provide an evaluation and analysis of the experience. The prompts were assigned at the end of class with the expectation that they would be submitted, along with all their other journal entries, at the end of the semester.

The core of the data comprises entries from 60 of the pre-service teachers' written journals, which were submitted electronically. The remaining nine pre-service teachers' journals were submitted in paper format and returned prior to data collection. Other data sources were a simple three-question pre-intervention survey asking whether the pre-service teachers had participated in timed drills as a student or had used them (or observed them being used) during their practicums and, finally, whether or not they expected to use timed drills in their future classroom. Notes were also taken of the in-class discussion that occurred prior to the intervention and of the activity debrief.

The data were coded and analysed using the methodology of modified analytic induction, which requires a phenomenon of interest and a working theory that can illuminate other similar situations (Bogdan & Biklen, 1998). It requires that data are coded and analysed for themes in order to develop or disconfirm the working theory (Gilgun, 1992). In this study the phenomenon of interest was the pre-service teachers' process of transition, and the theory began with the assumption that introducing a tension creates a consequence that can alter one's actions.

The themes were generated using NVivo analysis software. For example, for indicators of tension, we initially looked for utterances with emotional components such as mentions of anxiety, panic or anger. Noticing that these utterances frequently contained phrases like 'I never will' or 'I remember', we further divided the theme into tensions around future intentions and past memories. The latter underwent a subsequent iteration which resulted in subcategories of positive and negative memories.

Results and Analysis

In what follows, results from the survey as well as excerpts from the pre-service teachers' journals are used to exemplify the tensions that they are experiencing (or not) and how these tensions evolve as a result of the intervention. These results, and the accompanying discussions, are broken into the three salient stages of the pre-service teachers' evolution.

No Tension Between Intent and Action

Prior to participating in the intervention, 57 of the pre-service teachers ($n = 69$) indicated that they would likely be using timed drills in their future classrooms and 36 ($n = 69$) had used them during their practicums. When questioned, the majority felt timed drills were an effective way of learning basic facts. They expressed no tension between their intent and their action; this was an accepted practice that they fully anticipated utilizing in their classrooms. This is exemplified in the following two excerpts:

Julianne: I have grown up doing them [multiplication drills] and I don't see them as an issue.

Cate: I have memories of having to spew out math facts as fast as possible. I hated it but I think it's a good way to learn math facts.

In the journal entries, 17 of the pre-service teachers ($n = 60$) mentioned the enjoyment they experienced as young students participating in timed drills. They excelled at it and expressed positive emotions regarding the activity. It is not unexpected then that their initial surveys indicated that they would be using timed drills in their future classrooms. What was interesting were the 26 pre-service teachers who wrote about their negative experiences with timed drills as young students. Despite this, in their initial surveys, they too indicated their intention of using timed drills in their future classrooms. Their personal experiences were not enough to overcome their ingrained acceptance of this common yet pedagogically unsound teaching practice.

Creating a Tension Between Intent and Action

The intervention used to create a tension was a mock timed multiplication drill. Reliving the familiar experience of timed drills as an adult brought to bear not only intense anxiety but also all the negative feelings this type of intervention had caused them as a child. The journal entries contained vivid descriptions of the experience as seen in the following excerpts:

Cate: The minute you told us to stand up and that we will be doing multiplication questions, I went into a panic. My heart was racing, my stomach was clenching and I felt as if my brain was freezing.

Jennifer: It's definitely eye-opening, having that memory from almost 20 years ago and then the feeling of panic that I had when I thought that it was going to happen all over again in a university class.

Meryl: I came home thinking about all of those students who were in my own grade 3 class years ago that must have just been riddled with anxiety. There is something incredibly disturbing about that. Moreover, there is something even more disturbing that this is still a very, very commonly used practice. My own SA (mentor teacher) did it throughout my practicum.

Natalie: After you revealed that we actually weren't going to do this activity, and we debriefed it, I realized just how unhealthy it was for me to think that this was a normal way of teaching.

What emerged from the journals was that introducing the timed drill in an authentic manner was vital to the success of the intervention. Experiencing the activity as an adult learner highlighted the disconnect between their intent and their action. The excerpts reflect the recognition that the action of a timed drill interfered with their intent to have students master basic facts. In revealing the folkway of timed drills, we made room for doubt and uncertainty to creep into the pre-service teachers' mental image of timed drills. This emerging awareness can ultimately create what we call a useful tension in that it can lead to reflection and change in practice.

Consequence of the Tension on Action and Intent

The anxiety experienced by the pre-service teachers during the intervention was intense, and this was reflected in the journal entries where 47 of the pre-service teachers ($n = 60$) wrote about the negative effect they felt when asked to participate in a timed drill. Consequently, they were able to redirect this self-awareness to an understanding that children in their future classrooms would likely experience the same feelings—as we see in the following two excerpts:

Sandra: After we debriefed this activity, I realized how many people in our adult class felt uncomfortable with timed drills and being out on the spot in front of the rest of the class. This definitely was comparable, over even more, to the type of feelings and nerves we may see in our own classroom while teaching children. Only a small amount of students would love this activity, while the rest of the class would face nervousness, anxiety and worry.

Marion: When debriefing, I found it relieving and surprising to know how many other people felt the same way I did. Standing in a room full of adults who are becoming teachers, looking around at how much anxiety was caused by this one activity, I can only imagine in a room full of young students how they would feel.

The journal entries revealed that the intervention served as a means of reflection on the pre-service teachers' own future practice. The newfound pedagogical tension resulted in 51 of the pre-service teachers ($n = 60$) stating that they no longer felt that timed drills had a place in their classroom. As one pre-service teacher wrote:

Reese: As a teacher of mathematics, I will never force my students to do timed drills. After experiencing anxiety when you suggested we do this and seeing the anxiety it provoked in my peers, I was able to understand the anxiety that this causes in our students when we do the same to them.

Discussion and Conclusion

In answer to our research question, the shift from acceptance of an established teaching practice to a determination never to use it began with the introduction of what Berry (2007) describes as a tension between intent and action. This resulted in a consequence that disrupted the balance between the pairing, and we suggest that this will cause the pre-service teachers to seek out a new action.

It was readily apparent that, prior to the intervention, there was an absence of tension between the pre-service teachers intent to help students learn basic facts and the action of timed drills. They believed it to be an effective technique that would help them reach their goal. Considering that tension can be the impetus for change in practice (Lampert, 1985), a lack of tension is a strong indicator that the pre-service teachers would utilize timed drills in their future classrooms.

Participating in the intervention provided a new lived experience and created a tension that unseated the folkway of timed drills. In reflecting on that tension, the pre-service teachers realized that the action they were considering using to help students learn basic facts interfered with that very intention. The end result is that their initial action is no longer satisfactory for reaching their goals. The intent to have students learn their basic facts remains, but they will be searching for a new action to implement that will help them achieve that aim.

Berry's (2007) tension between intent and action offers a way of reframing traditional folkways of teaching and learning through reflection on experience. As pre-service teachers are unlikely to reflect on practices, which they view as common and accepted, it is the teacher educator who must devise a way to make these 'folkways' self-evident. Analysis of the data in this study revealed that purposefully creating a tension was useful in altering pre-service teachers' conceptions of timed drills. The intervention resulted in a useful tension which became a source of reflection and praxis.

References

Berlak, A., & Berlak, H. (1981). *The dilemmas of schooling*. London: Methuen.

Berry, A. (2007). *Tensions in teaching about teaching: Understanding practice as a teacher educator*. Dordrecht, The Netherlands: Springer.

Boaler, J. (2015). Fluency without fear: research evidence on the best ways to learn math facts. Retrieved from youcubed.org

Bogdan, R., & Biklen, S. (1998). *Qualitative research in education. An introduction to theory and methods*. Needham Heights, MA: Allyn & Bacon.

Buchmann, M. (1987). Teaching knowledge: The lights that teachers live by. *Oxford Review of Education, 13*(2), 151–164. https://doi.org/10.1080/0305498870130203

Gibbs, G. (1988). *Learning by doing: A guide to teaching and learning methods*. Oxford, England: Further Education Unit, Oxford Polytechnic.

Gilgun, J. (1992). Definitions, methods, and methodologies in qualitative family research. In J. F. Gilgun, K. Daly, & G. Handel (Eds.), *Qualitative Methods in Family Research*. Newbury Park, CA: Sage Publications.

Harper, N., & Daane, C. (1998). Causes and reduction of math anxiety in pre-service elementary teachers. *Action in Teacher Education, 19*(4), 29–38. http://dx.doi.org/10.1080/01626620.1998.10462889.

Kling, G., & Bay-Williams, J. (2014). Assessing basic fact fluency. *Teaching Children Mathematics, 20*(8), 488–497. http://dx.doi.org/10.5951/teacchilmath.20.8.0488.

Lampert, M. (1985). How do teachers manage to teach? Perspectives on problems in practice. *Harvard Educational Review, 55*(2), 178–195. http://dx.doi.org/10.17763/haer.55.2.56142234616x4352.

Liljedahl, P. (2014). Emotions as an orienting experience. In *Proceedings of the 20th International Conference on Mathematical Views*. Falun, Sweden: MAVI

Mason, J. (2002). *Researching your own practice: The discipline of noticing*. London: Routledge.

National Council of Teachers of Mathematics. (2014). *Procedural fluency in mathematics*. Reston, VA: NCTM.

Rouleau, A., & Liljedahl, P. (2015). Tensions in teaching mathematics: The case of Naomi. In *Proceedings of the 21st International Conference on Mathematical Views*. Milan, Italy: MAVI.

Tirosh, D., & Graeber, A. (2003). Challenging and changing mathematics teaching classroom practices. In A. Bishop, M. Clements, C. Keitel, J. Kilpatrick, & F. Leung (Eds.), *Second International Handbook of Mathematics Education* (pp. 643–688). Dordrecht, The Netherlands: Kluwer Academic Publishers. http://dx.doi.org/10.1007/978-94-010-0273-8_22.

University Teaching Assistants' Metaphors About Teachers' Role

Susanna Oksanen, Juulia Lahdenperä, and Johanna Rämö

Abstract According to Bullough (J Teach Educ 42(1), 43–51, 1991), metaphors can be viewed as a mirror of teachers' professional identity. This article reports what kind of metaphors university teaching assistants (TAs) at the Department of Mathematics and Statistics in the University of Helsinki, Finland, use for teacher's role. As a first phase of a longitudinal study, we analysed 35 TAs' metaphors using Beijaard, Verloop and Vermunt's (Teach Teach Educ 16, 749–764, 2000) model of teacher identity and metaphor manual for implementing this model (Löfström et al., Categorisation of teacher metaphors – Manual for implementing the Beijaard, Verloop & Vermunt Teacher Knowledge Base Model. Manual for NorBa project, 2011). Most of the TAs' metaphors were categorized as didactics expert or as self-referential. Also subcategories were analysed, and potential new subcategories found. The results also suggest that training can have an influence on the metaphors TAs use to describe their role as a TA.

Keywords Teaching assistants • Teacher professional identity • Teacher beliefs • Teacher role • Mathematics • Higher education

Introduction

Metaphors can be viewed as a mirror of teachers' professional identity (Bullough, 1991). The aim of the current study is to find out what kind of metaphors TAs use to describe their role as a TA.

In many universities, TAs have an important role in mathematics education. They often have more contact with the students than the lecturers of the mathematics courses. For example, in the University of Helsinki, every course has one or more

S. Oksanen (✉)
Department of Teacher Education, University of Helsinki, Helsinki, Finland
e-mail: susanna.oksanen@gmail.com

J. Lahdenperä • J. Rämö
Department of Mathematics and Statistics, University of Helsinki, Helsinki, Finland

H. Palmér, J. Skott (eds.), *Students' and Teachers' Values, Attitudes, Feelings and Beliefs in Mathematics Classrooms*,
https://doi.org/10.1007/978-3-319-70244-5_4

TAs who guide and support the students. Therefore, it is important to study TAs' beliefs and practices about teaching and learning mathematics. There is research on various aspects of university mathematics TAs' beliefs and practices (see, e.g. DeChenne, Enochs, & Needham, 2012; Ellis, 2014; Speer & Wagner, 2009), but there is no prior research on metaphors the TAs use for their role. In this study we examine TAs' beliefs and conceptions about their professional role expressed through metaphors. The aim of the study is to start a longitudinal research project on the TAs working at the Department of Mathematics and Statistics in the University of Helsinki. The results gained in this study enable us to plan and develop further research. The overall goal is to enhance the TA training and practices in the Department of Mathematics and Statistics.

Theoretical Background

Metaphors serve not only as a research instrument but also as an instructional strategy in teacher education; metaphors work as a tool when creating self-awareness and in-depth discussions of the nature of teachers' roles and their potential impact on students (Poom-Valickis et al., 2014).

Löfström, Hannula and Poom-Valickis (2010) concluded that metaphors can provide a fruitful starting point for exploring underlying beliefs and unconscious assumptions. Teachers' beliefs about mathematics and its learning and teaching are considered an indicator for certain behaviours in teaching. Richardson (1996) lists three categories of experience that influence knowledge and beliefs about teaching: personal influence, schooling and formal knowledge. Skott (2015) reports what aspects seem to have influence on beliefs according to previous research; personal life, practicum, schooling, work with colleagues, theoretical part of pre-service education and teacher development programme have an influence on the process of interpretation and construction on teachers' beliefs about mathematics, teaching and learning of mathematics and one self as a "mathematics person".

Tobin (1990) summarizes that beliefs about teaching and learning are associated with teaching roles, and metaphors are used to conceptualize these roles. A metaphor used to conceptualize a role can be changed in a process of changing the role, and new beliefs for a teaching role emerge when the role is reconceptualized.

Metaphor Categories

In the literature, there are two approaches for categorizing teacher metaphors: a data-driven approach assumes no a priori categories and builds the categories following the grounded theory approach; and the theory-driven approach uses a pre-existing system of categories and tries to categorize each metaphor into one of these.

The theory-driven approach was used by Löfström, Anspal, Hannula and Poom-Valickis (2010) when they studied metaphors about "the teacher". They based the categorization on Beijaard's et al.'s (2000) model of teacher identity according to which teachers' professional identity can be described in terms of *teacher as a subject matter expert*, *teacher as a pedagogical expert* and *teacher as a didactics expert*. Their results indicate that the model by Beijaard et al. (2000) can be applied as an analytical frame of reference when examining metaphors but that it would be useful to develop and expand the model further to include metaphors categorized as *self-referential* and *contextual metaphors.*

In this study we use the metaphor manual by Löfström et al. (2011) to analyse teachers' metaphors for their profession. We chose this model, because it has been tested by Oksanen and Hannula (2012) and Oksanen, Portaankorva-Koivisto and Hannula (2014). The categories are:

Teacher as subject expert This dimension of teacher identity highlights a profound knowledge base in his subject(s). Typical metaphors in the subject expert category describe the teacher as a source of knowledge, for example, *a book, a radio, and a computer.*

Teacher as didactics expert The teacher is a person who skilfully plans and manages learning process, as a person who knows how to teach specific subject-related content so as to support pupils' learning, for example, *a coach, a conductor, an engine, a road map, and a lighthouse.*

Teacher as pedagogical expert The teacher is seen as someone who supports the child's development as a human being. The understanding of human thought, behaviour and communication is an essential element in the teacher's pedagogical knowledge base, for example, *a mother, an older brother, and a firm tree.*

Self-referential metaphors These metaphors describe features or characteristics of the teacher's personality, with reference to the teacher's characteristics (self-referential) without reference to the role or task of the teacher, for example, *a machine, a candle, a sunshine,*and *a camel.*

Contextual metaphors These metaphors describe features or characteristics of the teacher's work or work environment or in other ways refer to characteristics of the environment (contextual). One might say that the metaphors describe where (physically, socially and organizationally) or in what kind of setting or environment the teacher works, for example, *a king, an actor, and a slave.*

Hybrids These metaphors include elements of more than just one of the above categories.

Unidentified Metaphors that could not be categorized in any of the categories presented above.

Teaching Assistants at the Department of Mathematics and Statistics

The Department of Mathematics and Statistics at the University of Helsinki is the biggest department in its field in Finland with over 1300 students. Typical undergraduate courses have 100–400 students. In autumn 2015 the department had 60 TAs who were either undergraduate students, master's degree students, doctoral students, or members of the staff.

The TAs have varied duties. Some TAs are affiliated with a lecture course and meet with a group of 20–30 students in a weekly tutorial. In the tutorials, problems solved by the students are discussed, and typically the students take turns in explaining their solutions on a blackboard. Other teaching assistants teach in drop-in sessions where the students can come and ask for help with any mathematical problems they have. Most of the tutorial and drop-in session TAs have a brief, voluntary training in the beginning of semester. In this study, these two types of TAs are referred as traditional TAs.

Since 2011, a fairly new teaching method, Extreme Apprenticeship (XA), has been used on many undergraduate courses. (For a detailed description of the method, see, e.g. Rämö, Oinonen & Vikberg, 2015.) In XA, the role of the TAs is to offer guidance to the students in a collaborative learning space where the students can spend as much time as they want. They lead the student subtly towards the discovery of a solution through a process of questioning and listening. Some of the weekly tasks are selected for inspection each week, and the TAs give written feedback on the students' solutions. During the course, the TAs go through a training by taking part in weekly meetings in which pedagogical aspects of their work are discussed. The recruitment process of the XA TAs includes an interview to ensure that they are interested in pedagogy and have motivation to teach.

Research Questions

- What kind of metaphors do university teaching assistants use for describing their role as a teacher?
- How do the metaphors given by traditional TAs and XA TAs differ?

Methodology

Instrument

The questionnaire for this longitudinal research concerning TAs practices and beliefs at the Department of Mathematics and Statistics was built in autumn 2015. As a part of this survey, TAs were asked to provide a metaphor characterizing the teacher's role.

The respondents were prompted with the beginning of a statement: "*As a teaching assistant I am like…*". They were also asked to add a brief explanation of their metaphor. This part was adapted from the questionnaire used in Nordic-Baltic Comparative Research in Mathematics Education. The TAs were also asked to give some background information concerning their academic experience and teaching experience.

Procedure and Sample

The data was collected during the feedback meeting of TAs in December 2015. There were 24 TAs present, and it took 30–45 min for them to fill in the questionnaire with tablets in the beginning of the meeting. The questionnaire was sent via e-mail to those TAs who were not present. In total, the questionnaire was given to 57 teaching assistants. The answer rate was 63%, giving $n = 36$. Of the respondents, 35 gave permission to use their answers.

Analyses

The analysis of the metaphors in the present study encompassed the following stages and actions:

1. The metaphor manual (Löfström et al., 2011) was read to guide the coding process; it consisted of explanations of categories and concrete examples of metaphors.
2. Two independent raters judged first the metaphor categories on a case-to-case basis. The metaphors and their explanations were analysed as a unit, as the metaphor itself may be used to express different meanings. The raters analysed the metaphors "from pure towards complex".
3. The codings of two independent raters were compared at the end.
4. In those cases where the metaphor was categorized completely identically, that category became the final category (65.5% of the cases, 23/35).
5. If the metaphors were coded partly identically and if the unit of analysis contained elements of two or more aspects, the one commonly used category used by both raters became the final category (26% of the cases, 9/35).
6. If two raters coded differently, a third rater was used, and when at least two coders agreed on coding, their coding was recorded (8.5% of the cases, 3/35).
7. If both raters used two or more same categories, these metaphors were classified as hybrids (8.5% of the cases, 3/35).
8. If the raters used different categories or the metaphor could not be identified in any category, these metaphors were removed (0%).

To analyse qualitative data and form subcategories, we used theoretical thematic analysis (Braun & Clarke, 2006, 2012). The stages in our analysis were (1) become familiar with the data, (2) generate initial codes, (3) search for themes, (4) review themes, (5) define and name themes and (6) produce the report. The theme analysis was carried out by one author of this paper, and the two authors of this paper compared the findings at the end.

Results

TAs' Background and Teaching Experience

There were 17 (49%) XA TAs and 18 (51%) traditional TAs. Of the XA TAs, 53% had mathematics as their major subject, and 47% were majoring in mathematics education. The majority of traditional TAs (72%) were mathematics majors.

The academic experience of traditional TAs (3 doctors, 6 doctoral students, 9 undergraduate or master's students) was more advanced when compared with XA TAs (1 doctoral student, 16 undergraduate or master's students). Traditional TAs had also more experience in teaching university mathematics. TAs' prior teaching experience is presented in detail in Table 1.

Metaphors

The most common metaphors used were categorized as didactics expert (40%) or as self-referential (38%); almost 80% of the metaphors were in either of these two categories. There were three metaphors in the pedagogical expert category and five hybrid metaphors that consisted of elements from two different metaphor categories. The categories included in hybrid metaphors were *subject expert*, *didactics expert*, *pedagogical expert* and *self-referential*. Hence, the *subject expert* category was present only in hybrid form. There were no metaphors in *contextual category*. The distribution of metaphor categories is presented in Table 2.

In the following subcategory analysis, the *hybrid* metaphors are included in both of their categories.

Metaphors describing teacher as *didactics experts* can be classified into two subcategories: *active* (10/17, 59%) and *passive* (7/17, 41%). The *active didactics expert* metaphors describe teachers who are present in the learning situation and are striving for better results both in teaching and learning (e.g. a *multifunction device. As a*

Table 1 The previous teaching experience of XA TAs and traditional TAs

		Previous university mathematics teaching experience				Other teaching experience
	#	None	<1 year	1–2 years	>2 years	
XA TA	17	35%	24%	41%	0%	71%
Traditional TA	18	17%	17%	39%	28%	28%

Table 2 The distribution of metaphors

n	Subject expert	Didactics expert	Pedagogical expert	Self-referential	Contextual	Hybrids
35	0	14 (40%)	3 (9%)	13 (37%)	0	5 (14%)

teaching assistant, I try to adapt to the student's way of thinking and the instruction situation; I approach the task in several different ways if I cannot get on the same wavelength with the student). The passive didactics expert metaphors describe teachers who are there to support the learners when needed (e.g. a *caretaker. I ensure that the student has the required mental state for learning, the necessary equipment, and a presence of support so that they can learn and find out by themselves*).

A closer analysis of the *self-referential* metaphors shows that there are five subcategories present: *lifelong learning* (4/16, 25%); *variability of mathematics teachers' job* (1/16, 6%); *persistence, bile or suitability to the job* (3/16, 19%); *mathematics teacher from student's perspective* (5/16, 31%); and *humour* (3/16, 19%). Two of these subcategories are new and not present in Oksanen et al. (2014), namely, *mathematics teacher from student's perspective* (e.g. *a gentle and wise bear. I might be a little scary, but then the students notice that I am a teddy bear. In addition, sometimes I do disservices by giving too much advice*) and *humour* (e.g. a*n analytic function. I obtain my maximum at the boundary*). On the other hand, there are no metaphors in *the big amount of work* subcategory, which, in contrary, was present in Oksanen et al. (2014).

The metaphor category distribution for traditional TAs and XA TAs is presented in Fig. 1. Traditional TAs gave more self-referential metaphors than XA TAs, and all metaphors in the pedagogical expert category were given by XA TAs.

In the hybrid category, the metaphors given by traditional TAs were from the categories subject content expert, didactics expert, pedagogical expert and self-referential. The hybrid metaphors given by XA TAs were from the categories subject content expert, didactics expert and pedagogical expert.

Discussion

In this study we asked TAs to fill in the following sentence: "As a teaching assistant I am like…". In 89% of the cases, the two independent raters used completely identical or partly identical categories. This result indicates that the metaphors are sometimes very complex and difficult to analyse.

When categorizing the metaphors, it was important to analyse not only the metaphor but also the provided explanation. However, this type of question might give a limited view of the TAs' beliefs of their role as a TA as they need to choose only one metaphor. On the other hand, the posing of the question could result in a more focused answer. The method needs to be further validated with interviews and focus group discussions in order to find out the nature of the methods' limitations.

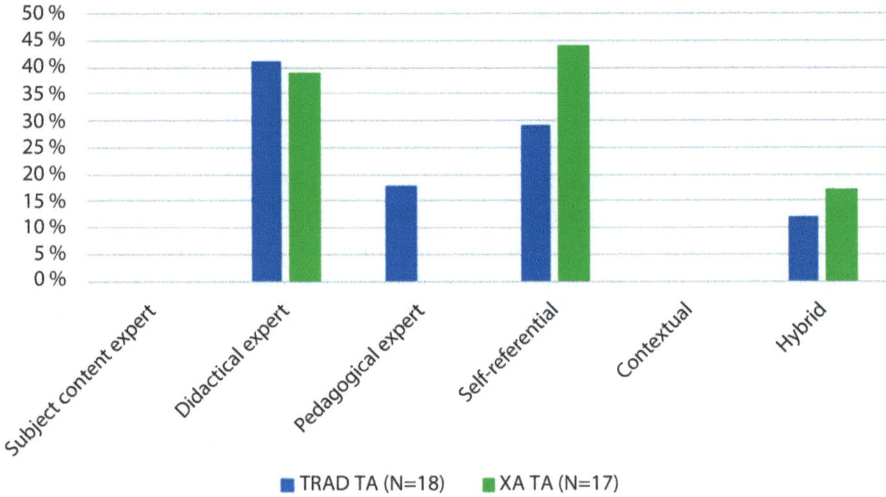

Fig. 1 Metaphor category distribution for traditional TAs and XA TAs

When looking at the TAs' metaphors, there were no metaphors found in category teacher as subject expert. This is surprising, as one would think that in university mathematics context, subject expertise would be emphasized by the TAs. As this research project started only last autumn, the sample was small ($n = 35$). Further research is needed to find out if TAs provide any metaphors in these categories or if the findings of this study were just a coincidence.

Another category that did not occur in this study was contextual metaphors. In previous studies, school teachers have provided contextual metaphors that describe their dissatisfaction with their job; they see their job too demanding or multifunctional (Oksanen et al., 2014). The metaphors in this sample do not suggest that the context where the TAs work would raise negative or positive feelings. One reason explaining this difference could be that the TAs' work is usually temporary, and they are not as engaged with it as school teachers. Therefore the problems rising from their work might not burden TAs as much as school teachers.

When looking deeper into the subcategories of the self-referential metaphors, the results indicate that the TAs' self-referential metaphors do not reflect much hesitation or doubt on their suitability to the job. This seems natural as for most of the TAs teaching is not their main job, and they do not need to be as committed as school teachers. When looking at the TAs self-referential metaphors and their subcategories, two new categories were found: mathematics teacher from student's perspective (31%) and humour (19%). Further research needs to be done to find out if any new subcategories appear.

When looking at the didactical metaphors, 41% of the TAs gave a passive didactical metaphor. Oksanen et al. (2014) report that pre-service teachers gave 37% and in-service teachers 30% of their didactical metaphors in a passive tense. This could be explained by the fact that 72% of the TAs are still undergraduate or master's students and don't have much experience in teaching.

There are some differences when it comes to the metaphors given by the traditional TAs and XA TAs. Traditional TAs give more self-referential metaphors, and XA TAs give more metaphors in the pedagogical expert category. These differences could be explained by the more intensive training the XA TAs receive, in which pedagogical aspects of their work are emphasized. Also, the XA TAs are interviewed before they are hired to ensure that they are interested in pedagogy and have motivation to teach. This can result in them giving more metaphors in the pedagogical expert category.

In previous studies, in-service teachers have given more metaphors in didactics expert and pedagogical expert categories than pre-service teachers (Oksanen et al., 2014). This suggests that these two categories are emphasized when a teacher gains more experience. In this light, it is interesting that in our study, the traditional TAs are more experienced than XA TAs, but they do not give more metaphors belonging to the didactical and pedagogical expert categories.

References

Beijaard, D., Verloop, N., & Vermunt, J. D. (2000). Teachers' perceptions of professional identity: An exploratory study from a personal knowledge perspective. *Teaching and Teacher Education, 16*, 749–764. https://doi.org/10.1016/S0742-051X(00)00023-8

Braun, V., & Clarke, V. (2006). Using thematic analysis in psychology. *Qualitative Research in Psychology, 3*, 77–101. https://doi.org/10.1191/1478088706qp063oa

Braun, V., & Clarke, V. (2012). Thematic analysis. In H. Cooper (Ed.), *APA handbook of research methods in psychology* (Vol. 2, pp. 57–71). American Psychological Association. https://doi.org/10.1037/13620-004

Bullough, R. V. J. (1991). Exploring personal teaching metaphors in pre-service teacher education. *Journal of Teacher Education, 42*(1), 43–51. https://doi.org/10.1177/002248719104200107

DeChenne, S., Enochs, L. G., & Needham, M. (2012). Science, technology, engineering, and mathematics graduate teaching assistants teaching self-efficacy. *Journal of the Scholarship of Teaching and Learning, 12*(4), 102–123.

Ellis, J. F. (2014). Preparing future college instructors: The role of graduate student teaching assistants (GTAs) in successful college calculus programs. Doctoral dissertation: University of California, San Diego.

Löfström, E., Anspal, T., Hannula, M. S., & Poom-Valickis, K. (2010). Metaphors about 'The Teacher': Gendered, discipline-specific and persistent? In J. Mikk, M. Veisson, & P. Luik (Eds.), *Teacher's personality and professionalism. Estonian studies in education* (pp. 105–122). Frankfurt am Main: Peter Lang Publishers House.

Löfström, E., Hannula, M. S., & Poom-Valickis, K. (2010). The cabbageworm in the classroom: Metaphors as expressions of students' beliefs about the teacher role. In M. F. Pinto & T. F. Kawasaki (Eds.), *Mathematics in different settings. Proceedings of the 34th annual Conference of the International Group for Mathematics Education* (Vol. 3, pp. 225–232). Belo Horizonte, Brazil: PME.

Löfström, E., Poom-Valickis, K., & Hannula, M. S. (2011). Categorisation of teacher metaphors – Manual for Implementing the Beijaard, Verloop & Vermunt Teacher Knowledge Base Model. Manual for NorBa project.

Oksanen, S., & Hannula, M. S. (2012). Finnish mathematics teachers' beliefs about their profession expressed through metaphors. In *Proceedings of the 18th Conference of the Mathematical Views*. Helsinki, Finland.

Oksanen, S., Portaankorva-Koivisto, P., & Hannula, M. S. (2014). Teacher metaphors – Differences between Finnish in-service and pre-service mathematics teachers. In *Proceedings of the 38th Conference of the Psychology of Mathematics Education*. Vancouver, Canada.

Poom-Valickis, K., Oder, T., Kislenko, K., Talts, L., Elvisto, T., & Lepik, M. (2014). Teachers' beliefs regarding their professional role. In A. Liimets & M. Veisson (Eds.), *Teachers and Youth in Educational Reality* (pp. 61–80). Frankfurt am Main: Peter Lang.

Rämö, J., Oinonen, L., & Vikberg, T. (2015). Extreme apprenticeship – Emphasising conceptual understanding in undergraduate mathematics. In K. Krainer & N. Vondrová (Eds.), *Proceedings of the Ninth Congress of the European Society for Research in Mathematics Education* (pp. 2242–2248). Prague, Europe: Charles University in Prague, Faculty of Education and ERME.

Richardson, V. (1996). The role of attitudes and beliefs in learning to teach. In J. Sikula (Ed.), *Handbook of research on teacher education* (pp. 102–119). New York: MacMillan.

Skott, J. (2015). From beliefs to dynamic affect systems in mathematics education: Exploring a mosaic of relationships and interactions. In B. Pepin & B. Roesken-Winter (Eds.), *Advances in mathematics education*. Springer. https://doi.org/10.1007/978-3-319-06808-4

Speer, N., & Wagner, J. (2009). Knowledge needed by a teacher to provide analytic scaffolding during undergraduate mathematics classroom discussions. *Journal for Research in Mathematics Education, 40*(5), 530–562.

Tobin, K. (1990). Changing metaphors and beliefs: A master switch for teaching? *Theory Into Practice, 29*(2), 122–127. https://doi.org/10.1080/00405849009543442

Prospective Teachers' Conceptions of the Concepts Mean, Median and Mode

Karin Kihlblom Landtblom

Abstract This paper examines the conceptions of mean, median and mode expressed by prospective teachers. A constant comparative method was used to analyse responses to a questionnaire. The results identified prospective teachers to express procedural knowledge rather than conceptual knowledge. Their descriptions resonate with definitions of the averages, with very few comments on how to teach average and statistical literacy. The results of this research have implications to inform essential course content in teaching statistics on teacher education programmes in the future.

Keywords Mean • Median • Mode • Prospective teachers • Comparative methods • Procedural knowledge • Conceptual knowledge • Average • Statistical literacy • Statistics

Introduction

There are few studies on teachers' understanding of statistics (Jacobbe & Carvahlo, 2011), but the results of these studies show that teachers', also prospective teachers, understanding appears not to be very different from students in school. Other studies indicate prospective teachers' understanding of statistical concepts to be procedural and consist of a collection of isolated rules rather than a conceptual scheme (Leavy, 2010). Jacobbe and Carvahlo (2011) suggested that the reason for this is that a more sophisticated level of knowledge about averages has not appeared in

K.K. Landtblom (✉)
Department of Mathematics and Science Education, Stockholm University, Stockholm, Sweden
e-mail: karin.landtblom@mnd.su.se

H. Palmér, J. Skott (eds.), *Students' and Teachers' Values, Attitudes, Feelings and Beliefs in Mathematics Classrooms*,
https://doi.org/10.1007/978-3-319-70244-5_5

teacher education systematically. This higher level of understanding, also known as statistical literacy, is something that teacher education needs to focus on (Ben-Zvi & Garfield, 2004; Shaughnessy, 2007). Statistical literacy implies that it is not enough to know only procedures; statistical literacy "involve(s) more than understanding the arithmetic mean" (Jacobbe & Carvahlo, 2011, p.207). Knowledge of averages should cover, for example, how to make sense of data as linking various means for average and averaging (Gal, 1995; Stack & Watson, 2010). This indicates that there is not only a gap that needs to be addressed in teachers' training on the concept of average but also a need for curricular goals to make clear how to teach averages (Gal, 1995). As indicated, statistical literacy also includes conceptual knowledge. Knowledge of a concept can be described as a scheme, as knowledge that develops through "networks consisting of connections between discrete bits of information about the measures that are formed" (Groth & Bergner, 2006, p.39). In this paper prospective teachers' conceptions of mean, median and mode are studied through the definition that conception is "a general notion or mental structure encompassing beliefs, meanings, concepts, propositions, rules, mental images and preferences" (Philipp, 2007, p.259).

One way to inform teacher education on prospective teachers' thinking is to analyse descriptions of prospective teachers' content knowledge (Groth & Bergner, 2006). In this study it is of interest to gain further knowledge about how prospective teachers locally understand statistical concepts. In the light of the above, the aim of this study is to study prospective teachers' conceptions or, at least in part, on their knowledge and understanding of mean, median and mode in relation to explaining these concepts to 4–6-year students. Thus, the research question for this study is: How do prospective teachers conceptualise the concepts of mean, median and mode to a student in years 4–6? The conceptions were analysed from a current group of prospective teachers' answers to a questionnaire.

Background

According to Hiebert and Lefevre (1986), knowledge can be thought of as conceptual or procedural. Conceptual knowledge as rich in relationships and procedural knowledge is made up of two parts: formal language (symbol representation system) of mathematics or of algorithms and rules (ibid, 1986). Research shows that students generally have procedural knowledge of isolated rules and reliance upon procedural algorithms rather than conceptual knowledge (Jacobbe & Carvahlo, 2011). Within statistics that is apparent when it comes to the concept of mean, students show ability to calculate the mean but show little basic conceptual understanding related to the concept (Cai & Moyer, 1995; Leavy & O'Loughlin, 2006). To achieve a broader understanding of statistical concepts, students need to understand and use statistical ideas at different levels. That includes competence and understanding of "the basic ideas, terms, and language of statistics" (Rumsey, 2002, s.2). Competences of statistics required for statistical literacy are brought up by Watson

(1997), who has summarised these skills in a three-tiered hierarchy: (a) a basic understanding of statistical terminology, (b) embedding of language and concepts in a wider context and (c) questioning of claims, with the aim to develop statistical literacy. Hence, both concepts and terminology are important components. To conceptualise averages, one need not focus only on calculation; there are also other intuitive important ideas concerning averages (Jacobbe & Carvahlo, 2011; Stack & Watson, 2010). Or as Gal (2000) puts it, doing statistics is not equivalent to understanding statistics.

Ideas about averages are discussed in various ways. There are at least three different perspectives on the term average: based on social experiences, based on media or based on the curriculum (Stack & Watson, 2010). Another way to approach students' understanding of average is to consider the following four perspectives: (a) average as modal, (b) average as what is reasonable, (c) average as the midpoint or (d) average as an algorithmic relationship (Rusell & Mokros, 1991). These different ideas are now exemplified below.

Averages

Traditionally in the teaching of averages, there has been a strong focus on the teaching of mean (Jacobbe & Carvahlo, 2011; Leavy & O'Loughlin, 2006). One reason for that could be that median is computationally more simple than the mean (Lesser, Wagler, & Abormegah, 2014) or in what kind of data is used (Mayén & Diaz, 2010).

Mean is often connected to an algorithm, a procedure, implying to add up and divide by the number of values, regardless of outliers (Jacobbe & Carvahlo, 2011). One way to bring an understanding to students would be to encourage explanation of how to find the mean and develop formulas that use everyday language, making clear connections as a useful life skill (Rumsey, 2002). Median is viewed as easier to understand than mean (Leavy & O'Loughlin, 2006), but still there are many aspects of the concept of importance, not least to a teacher (Lesser et al., 2014). One aspect is that median is not always a value in the dataset; another is difficulties when ordering data (Groth & Bergner, 2006). Importantly for teacher education, research informs us that elementary teachers, in particular, have several difficulties in determining medians in graphic data (Friel & Bright 1998), determining median from a set of unordered data (Zawojewski & Shaughnessy, 2000), ordering datasets and describing median as a centre of something but being unclear what something is (Mayén & Diaz, 2010). However, there is little research on how mode is conceptualised (Groth & Bergner, 2006). Mode is often described as the most frequent or the most popular in a dataset. At first in school, using only nominal data, it is easier to calculate the mode, but when numerical data appears, some confuse the variable value with the frequency (Watson, 2014). Understanding that there might be more than one mode and the importance of a mode, or not, is also important knowledge for teachers to understand (Watson, 2014).

If students are to choose between mean and median to describe a dataset, they often choose mean, without regard to distribution (Groth, 2013). One reason for not being able to choose could be if students "have been exposed to only non-contextual situations where the objective is to correctly perform a calculation" (Jacobbe, 2008). Another reason is if the students' knowledge has developed in isolation of one another (Jacobbe, 2008). One way to improve the students' intuitive understanding is to use real data instead of invented datasets (Stack & Watson, 2010). Working with real data and letting the students describe their choice would be one way to gain arguments and conceptual knowledge (Groth, 2013). Varying kinds of data could be of importance to contrast mean and median. Quantitative data seems prevalent than ordinal data when teaching averages and works for both mean and median. Some students, who do not perceive the difference, transform ordinal data to quantitative data and then calculate the mean. Working with both ordinal and quantitative data could be one way to contrast median and mean. This is a way to illuminate that you can decide a median with ordinal data, but not calculate a mean (Mayén & Diaz, 2010).

In sum, research informs us that teacher education programmes need to address several conceptions and misconceptions about averages. Teacher education needs to focus on specific knowledge development, e.g. why averages tell different things about a dataset, which averages are best to use under different conditions and why they do or do not represent a dataset (Garfield, 2002). One explanation suggests that procedural knowledge and calculation have a strong perception in statistics (Rumsey, 2002). To calculate an average is just the process to gain information; it does not demonstrate the ability to understand what average measures or how it is used (Rumsey, 2002).

This study could give an indication of prospective teachers' conceptual understanding and intuitional ideas of averages, thus providing indicators on what is the essential content in a teacher education course.

Methodology and Methods

This pilot study, of a single case (Yin, 2013), reports initial findings collected through a questionnaire in a teacher education course. The respondents are prospective teachers for school year 4–6 on their sixth semester of eight. The particular course in which the study was performed is the third course out of four dealing with teaching mathematics. The aim of the pilot was exploratory, using four open questions directly related to their conceptions on averages, three of which will be discussed here.

The questions were formulated as *how* questions, to identify any gaps in respondents' knowledge related to mean, median and mode. Such questions are appropriate in case study research according to Arthur, Waring, Coe, and Hedges (2012), as they offer possibilities to compare individuals' descriptions, definitions and understandings of conceptions, in this case, averages. The questions asked were: (1) How would you explain the concept *mean* to a student in years 4–6? (2) How would you explain the concept *median* to a student in years 4–6? (3) How would you explain the concept *mode* to a student in years 4–6? The questions are open in their character where the linguistic elements of how and explain and years 4–6 were provided to provide a structure and context in their answers towards a teaching situation.

The anonymity was an important condition for the prospective teachers to feel confident that their answers would not affect their grades in the course. This was confirmed in the questionnaire as well as orally declared in the course introduction. The response rate was 63% (29 out of 46).

The data has been analysed through a constant comparative method consisting of initial and selective coding (Glaser & Strauss, 1967). Initial coding implies staying close to the data and being open to what is going on in the data. Selective coding implies selecting the most frequent codes and how they relate to other codes identifying important relationships and differences (Arthur et al., 2012). The initial coding took place in several steps analysing the questions back and forth in order to identify what codes are to be found in the data. A starting point for the encoding was procedural and conceptual knowledge in accordance to Hiebert and Lefevre (1986) and competences of statistics required for statistical literacy (Watson, 1997). The final initial codes revealed the following: conceptual knowledge, procedural knowledge, context, colloquial concepts, usefulness, statistics (mathematics) and didactics (teaching). After grouping and comparing initial coding between the three averages, three tentative categories emerged: use of words, understanding averages and teaching explanation. These categories are a synthesis of what was seen as the result of this study. Alongside of the coding process, the codes and categories were read in conjunction to definitions of conceptual and procedural knowledge by Hiebert and Lefevre (1986) and skills required for statistical literacy by Watson (1997). The definitions the students used of the averages arithmetic mean, (referred to as mean in this text), median and mode were read in conjunction to the following definitions: *mean* the sum of the numbers divided by their quantity; *median* the middle of an odd number of observations and the mean of the two middle of an even number of observations; *mode* the observation value or observation values with the largest frequency (Kiselman & Mouwitz, 2008).

Results

The results will be presented and discussed out of the three categories that emerged in the analysis. In each category similarities and differences revealed between the participant's ideas about each average will be presented.

When comparing the initial codes for mean, median and mode, more varied combinations among the codes were found for mean. For median and mode, most of the respondents used definitions when explaining these averages. For mean, however, fewer definitions were used. The mean is more often explained in a context and with significantly more colloquial words. Another difference that emerged is that there are more words used for explaining mean and median than for explaining mode.

In Swedish the word *mean* is called *medelvärde* which is a composition of the two words *medel* (middle) and *värde* (value). This means that the word *medelvärde* signals a middle value. When describing what is measured or what is calculated, 15 used the word *value* and 9 used the word *number* or other synonyms indicating quantitative values. Sixteen used examples from a context, for example, age or weight. Another word often used was *genomsnitt* (used 29 times when describing the mean). The word *genomsnitt* is composed of the words *genom* (through) and *snitt* (cut), and the cut has many meanings. *Snitt* can also be used as shorthand for *genomsnitt* and was used three times. As a statistical term, it can describe average as well as *mean*, *median* or *mode*. It could also describe a typical variation or spread (NE, n.d.). In some of the answers, these different words are used in a way that could confuse. For example, in the following quote words, a variation of words is used to explain mean.

> *Medelvärdet* (the mean) is the same as *genomsnitt*. If one wants to find *genomsnittet* of, for instance, age in a family with four family members, then you add all the ages with each other and then divide with the number of family members – thus four. Then one will have *medelåldern* (the medium age) and *genomsnittsåldern* (the average age) of the family.

In Swedish, the word median does not have any particular synonym; it is natural to use the word median or to say something like the *middle observation*. The word median does not signal a value in the same way as *middle value* (mean), despite that when describing what is measured or what is calculated, 17 participants used the word value, 10 used the word number or other synonyms for number and finally 5 used examples from the context they had chosen, for instance, age or weight. Specifically for the case of median, the word *number line* is used seven times, and the word *number series* is also used seven times. The word *observation* is used just once. A typical description is "To calculate the median you line up the values in order of size, the smallest first and largest last. Then one looks up the number that is in the middle. That number is the median".

In Swedish the word *typvärde* is a composition of the two words *typ* (typical) and *värde* (value) and signals value, the same way as *medelvärde* (mean) does. Twenty-four used the word value, nine used the word number or other synonyms for number and finally seven used examples from the context they had chosen, for instance, age or weight. In particular for mode, the word frequency was used twice. The word observation is used one time. A typical description was "The mode is the value that appears most times. For example: 1, 1, 1, 1, 2, 1, 1. Here the mode is one".

How the students appear to understand averages is interpreted out of the data in different ways. One way is through the definitions and the other through numerical examples that are brought to some definition. Mean appears to be more familiar than median and mode. All but one definition on mean are correct, but when connecting the definition to a context, participants used it in an accurate way. In contrast the definitions of median and mode are incomplete or incorrect to a greater degree. Many show what happens to the median when having an odd number of observations but not for an even number, for example, "the value that is in the middle, neither largest nor least, but the middle". Seven respondents involved odd numbers in their definitions.

Understanding averages could also be seen through the code usefulness. Few students explained averages this way, but for those who did appear to show some kind of conceptual knowledge. Examples of answers in this code are "the mean is an average suitable to compare different observations"; "if you know that the mean is 10 years, then you know what activities could be suitable (e.g. for a party)"; "mean is interesting when you want to set an age on a group of people with various ages", or "the mean is 9.5. 9.5 is quite close to all the values, so that the mean we have calculated is a measure on the approximate price".

Some students compared mean and median in their explanations using outliers. The intention was to show that the median is more reliable in certain situations, for example, "important to choose an appealing and obvious example, for instance, five people's monthly salary where four people have about the same and one has twice the salary".

Using a context can be a way to provide an explanation into a teaching situation and/or to present data to be used in an explanation. All but one of these examples involved quantitative data. In the case of qualitative data, the frequency is confused with the variable. "The most frequent number in your series. For example, if you have 10 cars, 4 red, 3 blue, 2 white, 1 green, your mode is the most frequent number 4. It is the number that occurs most times". The context was predominantly used to complement a definition, rather than an example of how one can teach averages.

The teaching examples provided by the participants were varied. Two examples suggested using concrete material. One simply stated that this is a good idea when teaching mean, but the other offered a way to use it, e.g. "if we divide 16 in 4 piles, there will be 4 in each which means that the mean is 4". There were also a few examples on how data could be used in an explanation, for instance, to show that different data make the median a more appropriate average than mean, or the other way around, to be aware of ordering the data before deciding the median or to problematise mode by using data with more than one mode. Finally there were a couple of suggestions lacking argumentation. One asserted the importance to teach mean and median closely together in time, in order to teach the difference between the two concepts. Another stressed the importance of explaining the purpose of mode.

Discussion

The aim of this study was to investigate how this group of prospective teachers conceptualise the concepts mean, median and mode in relation to explaining these concepts to 4–6-year students. The result will now be discussed through conceptual and procedural knowledge according to Hiebert and Lefevre (1986) and competences of statistics required for statistical literacy (Watson, 1997). Finally the results will be discussed through Philipps'definition of conception (2007), written on page two in this paper.

The results show that a high proportion of the prospective teachers' explanations were predominantly related to definitions, rules or algorithms, of the averages investigated. This resonates strongly with Hiebert and Lefevre's (1986) definition of procedural knowledge. Any contexts used in their explanations are mainly descriptions on where numerical data can be found and show few examples of relationships within the concepts. It could be argued that expressing the definitions of averages can be equated to having a basic understanding of statistical terminology. However some of the participants' definitions are incomplete, and few embed the language and concepts in a wider context, the second point in Watsons' (1997) hierarchy. One reason for these results could be that there are many colloquial words used around the teaching and learning of averages, especially for mean. The results indicate how important both the concept and the terminology are when learning this concept, as many colloquial words or synonyms were used in the explanations here, just as Watson (1997) highlighted in her study. My conclusion is that words such as *medeltal* (middle number) and *genomsnitt* ("cut through" value), used proficially in this study by the participants as both mean and median, need to be understood and used in a very considered way.

As already mentioned, definitions were the main source to the explainations by the participants in this study. Yet trying to interpret their understanding, one cannot say more than if the definition is correct or not. The conclusion is that they show a strong procedural knowledge which Watson (1997) connects to the first level of understanding in her hierarchy of knowledge. Few prospective teachers showed any other understandings on mean and median than definitions. Mode seems to be more unfamiliar to the prospective teachers. This result is consistent with previous research (Groth & Bergner, 2006) and probably a result of their own schooling (Jacobbe & Carvahlo, 2011; Leavy & O'Loughlin, 2006). Altogether the results show few implications on understanding averages. A possible interpretation is that having only definitions, as a way to explain averages, does not give one the language to demonstrate conceptual knowledge. The conclusion is that working with different data levels of measurement and real data needs to be covered in teacher education on statistics in the future (c.f. Groth, 2013; Jacobbe, 2008; Mayén & Diaz, 2010; Stack & Watson, 2010).

The few examples provided by the participants were of two kinds: the first, brief instructions without argument and, second, more explicit examples of teaching. More explicit examples cover level 1 and 2 of Watsons' criteria (1997) for statistical literacy but also show some relationships within the concept which could be defined as conceptual knowledge according to Hiebert and Lefevre (1986). My conclusion is that the prospective teachers are not experienced in teaching averages, and the knowledge they have is not appropriate for doing this.

There are limitations in the study. The result presents only 29 prospective teachers' comments. The design of the questions affects the answers, the result and analyses. However there is a strong implication that the prospective teachers show mainly procedural knowledge and basic understanding of statistic terminology to a greater or lesser extent. The prospective teachers' conceptions about the concepts mean, median and mode can, in this pilot study, be described through Philipps' definition of concept (2007) as consisting of (I) concepts, mainly procedural; (II) rules, definitions of averages; (III) mental images, mainly definitions (few cases as didactic or mathematical); and (IV) beliefs and preferences, being a student rather than becoming a teacher. The knowledge the prospective teachers show is probably a result of their own schooling as they have not had any course yet at teacher education in teaching statistics. When comparing their knowledge to previous research, we can see that it is highly consistent. The implication of this is that there are three key aspects to consider in the future: (1) a need to strengthen prospective teachers' concept knowledge in order to develop statistical literacy; (2) a need to expand their network of knowledge within and between mean, median and mode; and (3) a need to teach the different data levels of measurement (e.g. nominal, ordinal, interval and ratio) in order to develop statistical literacy and gain a language to reason about statistics as a teacher. A result highlighted by this pilot study revealed the overuse of colloquial terminology used and thus need further investigation.

References

Arthur, J., Waring, M., Coe, R., & Hedges, L. V. (Eds.). (2012). *Research methods and methodologies in education*. Los Angeles: Sage.

Ben-Zvi, D., & Garfield, J. B. (2004). Statistical literacy, reasoning and thinking: Goals, definitions and challenges. In D. Ben-Zvi & J. B. Garfield (Eds.), *The challenge of developing statistical literacy, reasoning, and thinking* (pp. 3–15). Dordrecht, Netherlands/Boston, MA: Kluwer Academic Publishers. https://doi.org/10.1007/1-4020-2278-6

Cai, J., & Moyer, J. C. (1995). Middle School Students´ understanding of average: A problem solving approach. Paper presented at the Annual Meeting if the North American Chapter of the International Group for the Psychology of mathematics Education, Columbus, OH.

Friel, S. N., & Bright, G. W. (1998). Teach-stat: A model for professional development in data analysis and statistics for teachers K–6. In S. P. Lajoie (Ed.), *Reflections on statistics: Learning, teaching, and assessment in grades K-12* (pp. 89–117). Mahwah, NJ: Lawrence Erlbaum Associates.

Gal, I. (1995). Statistical tools and statistical literacy: The case of the average. *Teaching Statistics, 17*(3), 97–99. https://doi.org/10.1111/j.1467-9639.1995.tb00720.x

Gal, I. (Ed.). (2000). *Adult numeracy development: Theory, research, practice*. Cresskill, NJ: Hampton Press.

Garfield, J. (2002). The challenge of developing statistical reasoning. *Journal of Statistics Education, 10*(3), 58–69.

Glaser, B. G., & Strauss, A. L. (1967). *The discovery of grounded theory*. New York: Aldine.

Groth, R.E. (2013). Characterizing key developmental understandings and pedagogical powerful ideas within statistical knowledge for teaching framework. *Mathematical Thinking and Learning, 15*(2), 121–145. https://doi.org/10.1080/10986065.2013.770718

Groth, R. E., & Bergner, J. A. (2006). Preservice elementary teachers'conceptual and procedural knowledge of mean, median and mode. *Mathematical Thinking and Learning, 8*(1), 37–63. https://doi.org/10.1207/s15327833mtl0801_3

Hiebert, J., & Lefevre, P. (1986). Conceptual and procedural knowledge in mathematics: An introductory analysis. In J. Hiebert (Ed.), *Conceptual and procedural knowledge: The case of mathematics* (pp. 1–28). Hillsdale, NJ: Lawrence Erlbaum Associates, Inc.

Jacobbe, T. (2008). *Elementary school teachers' understanding of mean and median*. Proccedings of the ICMI Study 18 and 2008 IASE round table conference, Mexico.

Jacobbe, T., & Carvahlo, C. (2011). eachers understanding of averages. In C. Batnanero, G. Burrill, & C. Reading (Eds.), *Teacing statistics in school mathematics – Challenges for teaching and teacher education* (pp. 199–209). Dordrecht, The Netherlands: Springer. https://doi.org/10.1007/978-94-007-1131-0_21

Kiselman, C. O., & Mouwitz, L. (2008). *Matematiktermer för skolan*. (1. uppl.) Göteborg, Sweden: National Center for Mathematical Education (NCM), Göteborgs universitet.

Leavy, A. M. (2010). The challenge of preparing preservice teachers to teach informal inferential reasoning. *Statistics Education Research Journal, 9*(1), 46–67.

Leavy, A., & O'Loughlin, N. (2006). Preservice teachers understanding of the mean: Moving beyond the arithmetic average. *Journal of Mathematics Teacher Education, 9*(1), 53–90. https://doi.org/10.1007/s10857-006-9003-y

Lesser, L. M., Wagler, A. E., & Abormegah, P. (2014). Finding a happy median: Another balance representation for measures of center. *Journal of Statistics Education, 22*(3), 1–27.

Mayén, S., & Díaz, C. (2010). Median an easy concept? semiotic analysis of an open-ended task. Proceedings of the eighth international conference on teaching statistics. Voorburg, The Netherlands: International Statistical Institute. http://iase-web.org/documents/papers/icots8/ICOTS8_C265_MAYEN.pdf

Philipp, R. A. (2007). Mathematics teachers beliefs and affect. In F. K. Lester (Ed.), *Handbook on mathematics teaching and learning* (pp. 257–315). Charlotte, NC: Information age Publishing.

Rumsey, D. J. (2002). Statistical literacy as a goal for introductory statistics courses. *Journal of Statistics Education, 10*(3), 6–13.

Russell, S. J., & Mokros, J. R. (1991). What's typical? Children's ideas about average. In D. Vere-Jones (Ed.), *Proceedings of the third international conference on teaching statistics* (pp. 307–313). Voorburg, The Netherlands: International Statistical Institute.

Shaughnessy, J. M. (2007). Research on statistical learning and reasoning. In F. K. Lester (Ed.), *Second handbook of research on mathematics teaching and learning: A project of the NCTM* (Vol. 2, pp. 957–1009). Charlotte, NC: Information age Publishing.

Stack, S., & Watson, J. (2010). What's average? *AMT, 66*(3), 7–15.

Watson, J. M. (1997). Assessing statistical thinking using the media. In I. Gal & J. Garfield (Eds.), *The assessment challenge in statistics education* (pp. 107–121). Amsterdam, The Netherlands: IOS Press and International Statistical Institute.

Watson, J. (2014). What is typical for different kinds of data. *AMT, 70*(2), 33–40.

Yin, R. K. (2013). *Case study research: Design and methods*. Sage publications, Thousand oaks, California.

Zawojewski, J. S., & Shaughnessy, J. M. (2000). Take time for action: Mean and median: Are they really so easy? *Mathematics Teaching in the Middle School, 5*(7), 436–444.

Prospective Teachers' Approach to Reasoning and Proof: Affective and Cognitive Issues

Dorte Moeskær Larsen, Camilla Hellsten Østergaard, and Jeppe Skott

Abstract Reasoning and proof (R&P) are key elements in current reform efforts, but notorious for the problems they create for teachers. We present results from a pilot to an intervention study that seeks to alleviate these problems for prospective primary and lower secondary teachers in Denmark. The study introduces R&P in contexts that are "sufficiently close" to both academic mathematics and to instruction in school. The pilot asks, if this is a feasible approach. The part of the pilot presented here consists of responses by 57 prospective teachers to a qualitative questionnaire. The results show that many feel strongly about R&P, one way or another, but also that they have considerable problems with these processes to some extent irrespectively of their affective commitment. The results of the pilot confirm our approach for the main study.

Keywords Reasoning and proof • Pattern of participation • Teacher education • Affect • Emotions

Introduction

The interest in affect as it relates to mathematics teachers has been fueled by reform recommendations that change the relative emphasis in mathematics education from mathematical products (e.g. concepts, theorems, procedures) to processes (e.g. problem solving, communicating, representing). Indeed, the field is to a large extent a response to the rhetorical question of how one can expect *the reform* to materialize, if its priorities are not shared by the teachers. Further, the claim that alignment

---

D.M. Larsen (✉)
University of Southern Denmark, Odense, Denmark
e-mail: dmla@imada.sdu.dk

C.H. Østergaard
Metropolitan University College, Copenhagen, Denmark

J. Skott
Linnaeus University, Växjö, Sweden

H. Palmér, J. Skott (eds.), *Students' and Teachers' Values, Attitudes, Feelings and Beliefs in Mathematics Classrooms*,
https://doi.org/10.1007/978-3-319-70244-5_6

between teachers' affective relation with mathematics and the reform is necessary for reform impact is often supplemented with the claim that this is also sufficient. As Wilson and Cooney (2002) observe, "a considerable amount of research on teachers' beliefs [is] based on the assumption that what teachers believe is a significant determiner of what gets taught, how it gets taught, and what gets learned in the classroom" (p. 128).

We focus on prospective teachers' engagement with the mathematical process(es), of reasoning and proof (R&P). R&P, a sine qua non of mathematics, is promoted by current reform initiatives (e.g. Common Core State Standards, 2010) but acknowledged to cause affective and cognitive problems for students as well as for many of their teachers. Results from studies of Danish teachers substantiate this finding (e.g. Skott, 2013) and form the background to our intervention study that seeks to support prospective teachers in working with and teaching these processes.

We outline the background to the study (*R*easoning *a*nd *P*roving in *T*eacher *E*ducation, RaPiTE) and present the results of a qualitative questionnaire that prospective teachers completed as part of a pilot to RaPiTE. To understand if the approach of RaPiTE is feasible, we study the participants' affective and cognitive relationship with R&P as expressed in the context of the questionnaire. The results are that many prospective teachers express strong feelings about R&P but face considerable difficulties when engaging with these processes. More surprisingly, the difficulties are not only experienced by those, who explicitly lack confidence with mathematics. Prospective teachers who claim to be strong in mathematics, to enjoy doing it, and say that R&P are important in school have considerable problems with simple R&P tasks.

Conceptual Framework

In RaPiTE we use a framework called Patterns of Participation (PoP) (Skott, 2015a, 2015b). We do not focus on beliefs, understood as reified, mental constructs, but on teachers' (re-)engagement in mathematical, educational, and other social practices and discourses that in belief research are considered the basis of the reifications.

PoP draws on the notions of practice (Wenger, 1998) and figured worlds (Holland, Skinner, Lachicotte, & Cain, 1998) in social practice theory and of self in symbolic interactionism (SI) (Blumer, 1969; Mead, 1934). In social practice theory, practice is a communal endeavour, the meaning of which is negotiated as people engage with one another in communities that emerge in interaction. Figured worlds (FWs) are "collective as-if worlds in which particular characters and actors are recognised, significance is assigned to certain acts, and particular outcomes are valued over others" (Holland et al., 1998, p. 52). It is apparent that the two constructs share their emphasis on participation in socially constructed enterprises and the social constitution of human experiences, but FW is a broader concept than the one of practice and does not necessarily require substantial joint engagement among participants.

While the social construction of experience is fundamental in PoP, we seek to recentre the individual in participatory accounts of human functioning. We find symbolic interactionism (SI) helpful for this purpose. In SI terms, people take the attitude to themselves of individual or generalised others in interaction. Experiencing or envisaging the reactions of others, they instantaneously modify their contribution to the interaction based on previous experiences. For example, a teacher may foresee the reaction of students, who have previously challenged her professional competence and seek to position herself as a qualified mathematician. In other situations, however, she may take the attitude of weak and vulnerable students, who seek to remain part of the classroom community. In this situation, she may engage in a figured world that emphasizes broader pedagogical concerns than the students' mathematical learning.

From a PoP perspective, discrepancies between teachers' "beliefs" and classroom practice are no surprise. In fact, there is as much to explain, if there is apparent compatibility between teachers' discursive engagement with a reform initiative in a research interview and their contributions to classroom practice, as when this is not the case. In both cases, the analysis should document how the teacher's attempts to relate sensibly to the practices that unfold at the instant are informed by her re-engagement in a range of other ones, e.g. mathematics itself, collaboration with her colleagues, experiences from her teacher education programme, and many more.

Background and Rationale of RaPiTE

The reform literature on R&P suggests that students engage in a cycle of conducting explorations, making conjectures, and developing justifications (NCTM, 2008). This involves making generalisations and arguing for their truth value (Stylianides, Stylianides, & Shilling-Traina, 2013). The notion of proof refers to particular versions of the last element in the R&P cycle, namely, to justifications that (1) are based on previously accepted statements, (2) employ accepted forms of argumentation, and (3) use accepted modes of communication (Stylianides, 2007). Here, "acceptance" refers to two different practices, that of mathematics and that of the classroom. Empirical arguments, for instance, which students and teachers may accept do not qualify as proof, whereas generic (Rowland, 2002) or single-case-key-idea-inductive arguments (Morris, 2007) often do.

In RaPiTE we adopt a similar perspective on R&P. We build on the literature on teachers' relation to R&P (e.g. Bieda, 2010; Furinghetti & Morselli, 2011; Knuth, 2002; Stylianides et al., 2013) and on some of our previous studies of beginning teachers. Our studies did not initially focus on R&P, but results suggest that research participants' support to their students' engagement with R&P are rarely in line with reform recommendations. Often their students engaged in explorations and made conjectures, but they were rarely required to provide justifications, and if so modes of reasoning that do not qualify as mathematical were often accepted (Skott, 2009; Skott, Larsen, & Østergaard, 2011). In line with others (e.g. Gellert, 2000; Sztajn,

2003), we found that teachers were often concerned with other issues than their students' learning. Beyond that, a PoP interpretation of the interactions suggests that often teachers do not have enough relevant experiences with R&P to draw on in instruction. "Relevance" means close to both the mathematical practice of R&P *and* to instruction.

RaPiTE, therefore, seeks to avoid the two extremes of focusing either on school mathematics or on academic mathematics in mathematics teacher education. We do not seek to reduce the emphasis on either, but to develop prospective teachers' professional competence by engaging them in the mathematical practices of R&P in relation to mathematical tasks and contents that have been or may be used in school classrooms (Skott, in press). This includes balancing *proving why*, often using generic arguments (Rowland, 2002), and *proving that*. Exemplary tasks include:

1. *Prove or disprove (a) that the difference between two consecutive perfect squares is the sum of their bases and (b) that 8 divides n^2-1, if n is an odd integer* (from a grade 5 working with perfect squares, cf. Skott (in press)).
2. *30 = 6 + 7 + 8 + 9; 31 = 15 + 16; 32 = ...; 35 = 17 + 18 = 5 + 6 + 7 + 8 + 9 = ... What positive integers are the sum of other consecutive, positive integers* (from a reversal of the question of how to find the sum of the first *n* positive integers).

Beginning with tasks or conjectures used or developed in school (e.g. (1)), or with extensions of such tasks (e.g. (2)), RaPiTE seeks to address both the cognitive and affective challenges that primary and lower secondary teachers often have with R&P in their classrooms. The primarily cognitive ones include not distinguishing between mathematical and other justifications and not being sufficiently familiar with *proving that* to decide whether a students' conjecture is right or wrong. The affective ones include not prioritising that students learn about R&P, not recognising R&P as support to understanding the contents, and being insecure when working with R&P.

The part of the pilot dealt with below deals with the question of whether it is necessary to deal with these challenges in Danish teacher education and in particular how prospective teachers react to issues of R&P in the context of the questionnaire. Clearly, PoP does not interpret the results of a questionnaire as indicators of how teachers engage with R&P when teaching. The more modest question is how they engage discursively with educational issues related to R&P in the context of the questionnaire. It is a question for the main study, if and how they draw on these discursive practices in the teacher education programme, including their practicum.

Methods

The data for this paper consists of responses by 57 prospective mathematics teachers to a questionnaire concerned with mathematics education in general and with R&P in particular. The respondents are from two cohorts of prospective teachers, who enrolled at a fairly prestigious college in a major city in Denmark 2014 and 2015.

The college educates teachers for grades 1–9. All students study three school subjects, one of which must be Danish, English, or mathematics. The 57 respondents are all to qualify to teach grades 4–9 and are among a third of the students, who choose to study mathematics. They have all performed reasonably well in upper secondary school, and they have all passed mathematics at least at *level B*, the second highest level offered. Nineteen students have enrolled in a special programme, *Advanced Science Teacher Education* (ASTE), in which the other school subjects are all sciences. In general, we expect the respondents (not only ASTE students) to be better qualified and more interested in mathematics than the majority of Danish teachers for these levels.

The prospective teachers received the questionnaire on the day of their first class in mathematics at the college and handed it in within the next few hours. The questionnaire invites the participants to account for their prior experiences with mathematics, their reasons for becoming mathematics teachers, and their approach to and proficiency with mathematical reasoning. Most items are open, some asking participants to describe their general experiences with R&P or the role of R&P in school mathematics. Two other items are more specific and introduce a mathematical idea in a classroom context (Ball, 1988). One of them asks how the respondents think they would react to a grade 5 student, who proudly presents the conjecture that if the perimeter of a rectangle increases, so does the area; another item asks how to prove that the sum of two odd numbers is even, and how this result may be explained to school students and to the participants' fellow students in the teacher education programme.

The codes and categories were developed using methods inspired by grounded theory (Charmaz, 2006). Initially we identified sections of data that highlighted (1) the participants' emotional relation to R&P and (2) their perspective on R&P in mathematics education. Subsequently three and five codes were developed for these two dimensions, respectively. For our present purposes, we do not do a cross tabulation of these results, but present them along the two dimensions in turn and relate them to the participants' response to the two items on R&P in classroom contexts. In these two items, we were primarily interested in whether the research participants were able to relate to the tasks in a way that from our perspective was mathematically valid. The primary codes for these items were just *(almost) acceptable* and *unacceptable*.

Two points must be made in relation to our reading of the responses to the questionnaire. First our analysis is not based on the assumption that the participants' use of terminology corresponds to the meaning we attach to the terms and, for instance, that their use of *mathematical reasoning* resembles the meaning of the term outlined previously. One intention with the questionnaire was exactly to understand how the participants, formally among the best qualified prospective teachers in Denmark, relate to the notion of mathematical reasoning, affectively as well as cognitively, in a questionnaire at the beginning of their teacher education programme.

Second, and to reiterate, we do not assume that the responses to the questionnaire reflect stable, mental entities that necessarily align with how the participants react in other situations, e.g. in a classroom. For instance, they may respond more positively in the questionnaire to items on their affective relationships with mathematics, than if they had not just enrolled in a programme for mathematics teacher education. However, we also suggest that responding to a questionnaire at this particular time constitutes a situation so focussed on mathematics that it allows interpretations like the following: if the participants do not engage proficiently with R&P in the questionnaire, they are unlikely to do so in the much less structured situation of classroom interaction.

Results

The prospective teachers generally claim to be fond of mathematics, 46 of them using positive wordings (e.g. "I am a big fan of mathematics"). The general passion for mathematics is often combined with mathematical self-confidence, and 18 participants say that they were always good in mathematics, while 6 others claim to have had difficulties with the subject at times but now enjoy doing it. Two participants acknowledge having problems with the subject (e.g. "In fact I think mathematics is difficult"). The remaining participants did not indicate how qualified they consider themselves in mathematics. Surprisingly, the responses to the general affective items do not correspond well with the answers to the question of the participants' prior educational experiences. Only 15 claim to have had mainly positive experiences, 38 say that their experiences are mixed, and 4 say that they have had decidedly negative experiences with mathematics. In spite of this, few indicate that their relationship with mathematics is now in any way problematic.

The three codes on the emotional dimension of R&P are (1) primarily positive, (2) neutral or balanced, and (3) primarily negative. These codes are mainly constructed from responses of the participants' general experiences with R&P. Four of the codes on the role of R&P in schools are (1) (potentially) important for meaning-making and understanding; (2) something to be learnt by heart; (3) too difficult, convoluted, and/or unnecessary; and (4) a way of showing how you solve tasks. Finally some responses were coded as (5) explicitly acknowledging not to know what mathematical reasoning and proof are. These five codes are based both on responses to the question of the role of R&P in school mathematics and to questions on general experiences with R&P.

Among the affective responses, ten stood out as being almost purely emotional and not refer to any role for reasoning in education. Four of these responses are positive as in "I felt good about it" or "I loved it", while six are negative, for instance, "I hated it" or "I hate proofs and reasoning, because in upper secondary school there was no time to explain it properly […]", and "Proofs are the devil's making until you finally get it". All in all we coded 15 responses as primarily positive, 22 as neutral or balanced, and 9 as primarily negative. The remaining participants did not respond to the item or did so in ways that were uninterpretable along the affective dimension.

Forty-seven responses are not (only) emotional but affective in a broader sense and deal with the role of R&P in education. There are five categories of such responses:

1. Reasoning and proof as (potentially) important for developing understandings, providing meaning, and justifying results (18 participants), e.g.:
   - "It gives [you] a good understanding of mathematics, if you understand it, but it may take some time".
   - "When I began learning proofs I found it very interesting. To know how things are connected helped me understand".
   - "It was challenging. Often you had to keep a cool head […]. It was fun, because it was more difficult and you can understand why you do what you do".
2. Reasoning and proof are to be learnt by heart (5 participants), e.g.:
   - "Proofs are just something you learn by heart instead of finding out on your own".
   - "It is difficult […]. I learnt things by heart, and I was best [at it] when I just knew how to do it".
3. Reasoning and proof are difficult, convoluted, and/or unnecessary (7 participants), e.g.:
   - "Proofs may become very complicated, if they are used to prove things that are very remote from reality".
   - "If I get or have the tools to solve a task, it does not mean much to me that there is a proof behind it".
4. Reasoning and proof function as a way of explaining how you solve tasks (14 participants), e.g.:
   - "In grade 5 the new mathematics teacher wanted us to write down all the calculations – I was not pleased about it at the time. In upper secondary school it became far more essential to explain the calculations".
   - "In primary and lower secondary school it was no problem. I knew why and how to solve math problems".
   - "From grade 8 it was important to describe why a result was what it was, but only in word problems, but with them the result was only 1/3 [of the grade], the explanation was the rest".

5. Explicitly acknowledging not knowing what reasoning and proof are (3 participants), e.g.:
   - "I am way too untrained in proofs, so I have to work to get better".
   - "[I] do not know what mathematical reasoning is, but I suppose it is those samples that you make to see if a proof is right".

Twenty-six prospective teachers provide an acceptable response to the item about a student, who presents a false conjecture about the connection between the perimeter and area of a rectangle (e.g. pretending to talk to a student "Try to compare a 3×3 square with a 6×1 rectangle. [...] Is this interesting, different examples give different results?"). Thirty participants think the student is right (e.g. "This is completely right. Every time you make one side longer the area gets bigger [...] the perimeter and the area automatically grow bigger together"). One participant did not respond to the task.

The research participants found it even more difficult to prove that the sum of two odd numbers is even. Fifty-four of them responded, 12 providing an (almost) acceptable answer:

- "$2n + 1 + 2n + 1 <=> 4n + 2$ for all n in N. Otherwise one can prove by induction (not enough space here)".
- "It is in the definition of an odd number that it has remainder of 1 when divided by 2. Two odd numbers with the remainder of 1 that are added have the remainder of 2 -> no remainder".
- "The student has made a drawing of two rows of 3 and 5 counters, taking one counter from each and adding them. Arguing that now the rods both have an even number of counters, and adding 2 still makes the result even".

Three responses coded as (almost) acceptable are correct, if one accepts as a premise that the sum of two even numbers is even. In two of the cases, this is made explicit, e.g.:

- "An odd is always an even number + 1. B and A are even numbers, we know that when A and B are even, then $A + B = C$, C being even. Then $(A + 1)$ and $(B + 1)$ both odd. $A + 1 + B + 1 = C + 2$, C is even and 2 is even. Qed".

Talking to students, one participant provides an (almost) acceptable answer in the form of a single-case-key-idea-inductive argument. She makes a drawing (Fig. 1) and explains, "Because one number does not have someone to hold hands with, it gets one from the other, who also does not have a partner".

Among the 42 responses that we consider unacceptable, 11 give empirical arguments:

- "Adding two odd numbers we will always get an even number, e.g. $9 + 9 = 18$, $11 + 11 = 22$, $17 + 17 = 34$ / cannot ...".
- "You can see it if you just try a sufficient number of times that the result is always even. And it does not matter how big the numbers are. [...]".

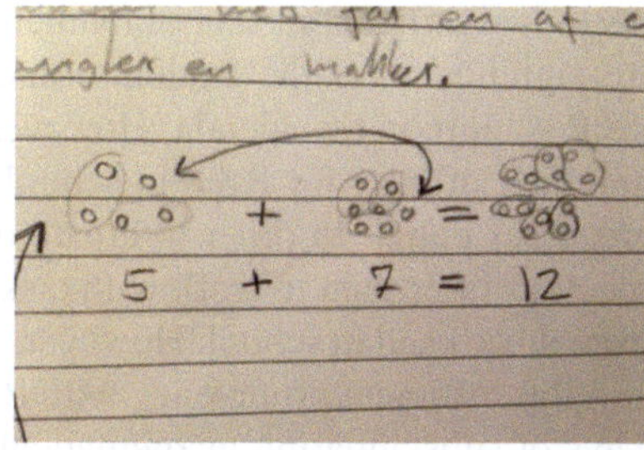

Fig. 1 The sum of two odds is even

- "I would try with some examples. 1 + 1 = 2 3 + 3 = 6 7 + 7 = 14. The same for some even numbers. [...] Okay, I have no idea how to prove this. (I am one of those students who never asked why, but just accepted because it made sense to me)".

In the last example, the student knows that his empirical argument does not suffice. This is the case for three of the students, who provide empirical justifications.

Other unacceptable responses are not easily categorisable. Some restate the result to be proved in other terminology (e.g. "Because the result if divided by 2 is a whole number"). Others introduce misinterpreted mathematical terminology that is unconnected to the question (e.g. "Tell them about the hierarchy of –, + and division. Tell them that when they have opposite signs, they cancel [...]"). And some say that proofs have no role in mathematics (e.g. "A claim may be substantiated, if it cannot be refuted. However, this is NEVER a proof. Mathematics cannot be proved").

Of the 46 prospective teachers who consider themselves good at mathematics and/or are fond of the subject, 18 provide a mathematically (almost) acceptable response to the perimeter-and-area item, and 8 do so to the item on the sum of two odd numbers.

Discussion and Conclusion

In the context of the questionnaire, most participants find it difficult to engage in R&P processes that qualify as mathematical. It is, considering other studies, unsurprising that many cannot make a formal proof that the sum of two odd numbers is even, and it is only to be expected that some give a few examples to support the result but admit to not knowing how to prove it. We find it more surprising that most respondents endorse invalid mathematical arguments. The participants may be considered a critical group of prospective teachers and the occasion of the questionnaire a somewhat critical situation (cf. the methods section). Yet, approximately 75% of those who are fond of and/or consider themselves good at mathematics accept invalid arguments.

The suggestion in RaPiTE is that R&P as taught in teacher education should be "sufficiently" close to both the discipline of mathematics and to the subject as taught in school. The pilot supports this idea. Some respondents are reluctant to accept

R&P as significant in school mathematics, but most do and many suggest that they are significant for the students' understanding of the contents. Affectively, then, R&P is high on the agenda. However, in the questionnaire, few of them are able to link their affinity for R&P in school contexts with an understanding of what these processes entail. In order to create or strengthen such links, it is, from a PoP perspective, important to work with tasks and conjectures that are or may be developed from those used in school. However, it is equally important that prospective teachers develop sufficient proficiency with R&P as a decidedly academic activity, including ways of substantiating mathematical claims. Otherwise they have no choice but to accept or reject student conjectures on empirical grounds, making their own and their students' involvement in the last of part of the R&P cycle illusory.

References

Ball, D. L. (1988). Knowledge and reasoning in mathematical pedagogy: examining what prospective teachers bring to teacher education. (Unpublished PhD dissertation), Michigan State University, East Lansing.

Bieda, K. N. (2010). Enacting proof-related tasks in middle school mathematics: Challenges and opportunities. *Journal for Research in Mathematics Education, 41*(4), 351–382.

Blumer, H. (1969). *Symbolic interactionism. Perspective and method.* Berkeley, CA: University of Los Angeles Press.

Charmaz, K. (2006). Constructing grounded theory. A practical guide through qualitative analysis. London: Sage.

Common Core State Standards. (2010). *Common core state standards for mathematics*. Retrieved from http://www.corestandards.org/wp-content/uploads/Math_Standards.pdf

Furinghetti, F., & Morselli, F. (2011). Beliefs and beyond: Hows and whys in the teaching of proof. *ZDM - The International Journal on Mathematics Education, 43*(4), 587–599. https://doi.org/10.1007/s11858-011-0316-7

Gellert, U. (2000). Mathematics instruction in a safe space: Prospective elementary teachers' views of mathematics education. *Journal of Mathematics Teacher Education, 3*(3), 251–270.

Holland, D., Skinner, D., Lachicotte, W., Jr., & Cain, C. (1998). *Identity and agency in cultural worlds*. Cambridge, MA: Harvard University Press.

Knuth, E. J. (2002). Teachers' conceptions of proof in the context of secondary school mathematics. *Journal of Mathematics Teacher Education, 5*(2), 61–88.

Mead, G. H. (1934). In C. W. Morris (Ed.), *Mind, self, and society from the standpoint of a social behaviorist*. Chicago: University of Chicago.

Morris, A. K. (2007). Factors affecting pre-service teachers' evaluations of the validity of students' mathematical arguments in classroom contexts. *Cognition and Instruction, 25*(4), 479–522.

NCTM. (2008). *Navigating through reasoning and proof in grades 9–12*. Reston, VA: NCTM.

Rowland, T. (2002). Generic proofs in number theory. In S. R. Campbell & R. Zazkis (Eds.), *Teaching and learning number theory* (pp. 157–184). Westport, CT: Ablex.

Skott, J. (2009). Contextualising the notion of belief enactment. *Journal of Mathematics Teacher Education, 12*(1), 27–46. https://doi.org/10.1007/s10857-008-9093-9

Skott, J. (2013). Understanding the role of the teacher in emerging classroom practices: Searching for patterns of participation. *ZDM - The International Journal on Mathematics Education, 45*(4), 547–559. https://doi.org/10.1007/s11858-013-0500-z

Skott, J. (2015a). Beliefs and brownies: In search of a new identity for 'belief' research. In L. Sumpter (Ed.), *Current state of research on mathematical beliefs XX. Proceedings of the*

MAVI-20 conference (pp. 33–45). Falun, Sweden: Högskolan Dalarna. http://lnu.diva-portal.org/smash/get/diva2:915269/FULLTEXT01.pdf.

Skott, J. (2015b). Towards a participatory approach to 'beliefs' in mathematics education. In B. Pepin & B. Rösken (Eds.), *From beliefs to dynamic affect systems in mathematics education. Exploring a mosaic of relationships and interactions* (pp. 3–23). Cham, Switzerland: Springer.

Skott, J. (in press). The goldilocks principle revisited: Understanding and supporting teachers' proficiency with reasoning and proof. In P. Błaszczyk, B. Pieronkiewicz, & M. Samborska (Eds.), *Mathematical transgressions*. Kraków, Poland: PWN.

Skott, J., Larsen, D. M., & Østergaard, C. H. (2011). From beliefs to patterns of participation: Shifting the research perspective on teachers. *Nordic Studies in Mathematics Education, 16*(1–2), 29–55.

Stylianides, A. J. (2007). Proof and proving in school mathematics. *Journal for Research in Mathematics Education, 38*(3), 289–321.

Stylianides, G. J., Stylianides, A. J., & Shilling-Traina, L. (2013). Prospective teaches' challenges in teaching reasoning-and-proving. *International Journal of Science and Mathematics Education, 11*(6), 1463–1490.

Sztajn, P. (2003). Adapting reform ideas in different mathematics classrooms: Beliefs beyond mathematics. *Journal of Mathematics Teacher Education, 6*(1), 53–75.

Wenger, E. (1998). *Communities of practice. Learning, meaning, and identity*. Cambridge, UK: Cambridge University Press.

Wilson, M., & Cooney, T. J. (2002). Mathematics teacher change and development. The role of beliefs. In G. C. Leder, E. Pehkonen, & G. Törner (Eds.), *Beliefs: A hidden variable in mathematics education?* (pp. 127–148). Dordrecht, The Netherlands: Kluwer.

Using Cases and Events in Teacher Education: Prospective Teachers' Preferences

Pessia Tsamir, Dina Tirosh, Esther Levenson, and Ruthi Barkai

Abstract The use of cases as a pedagogical tool in teacher education is seen as one way of bringing practice closer to theory. This study describes the use of cases in a university course for secondary school prospective mathematics teachers. The study investigates participants' views of cases taken from different sources and presented in different situations. In general, participants felt that the use of cases had an impact on their understanding of common mathematical errors but that cases based on mistakes they had themselves made during homework assignments were most meaningful.

Keywords Cases • Prospective secondary teachers • Preferences • Mathematical errors • Views

Introduction

Preparing future mathematics teachers is a complex process involving both academic and practical elements. Academic elements often include university or college courses aimed at promoting prospective teachers' mathematics knowledge needed for teaching (Ball, Thames, & Phelps, 2008). Practical elements often include fieldwork such as classroom observations and student teaching where the aim is to practice and apply what was learned in theory. Yet, bridging out-of-university practice and within-university academic courses can be a challenge (Zeichner, 2010). Some educators argue that clinical experiences should be central to teacher education and that all teacher preparation should stem from those experiences (Ball & Forzani, 2009). Others point out that prospective teachers' lack of experience may limit their observational skills and consequently limit what may be learned from fieldwork (Masinglia & Doerr, 2002). While we agree that clinical experiences are essential to teacher preparation, additional tools, such as analyzing classroom videos and cases, may also assist in bringing the classroom practice

P. Tsamir (✉) • D. Tirosh • E. Levenson • R. Barkai
Tel Aviv University, Tel Aviv, Israel
e-mail: pessia@tauex.tau.ac.il

H. Palmér, J. Skott (eds.), *Students' and Teachers' Values, Attitudes, Feelings and Beliefs in Mathematics Classrooms*,
https://doi.org/10.1007/978-3-319-70244-5_7

closer to future teachers, while enhancing future teachers' mathematics knowledge and pedagogical content knowledge (e.g., Markovits & Smith, 2008; Santagata & Guarino, 2011).

In this paper we describe a university course which, among other tools, used cases as a tool for promoting participants' mathematics knowledge, as well as their knowledge of common mathematical errors made by secondary school students. We use the term case in a broad sense to mean an actual event that occurred at some time in some learning situation and which may be generalized beyond the specific event to a larger set of mathematics education ideas. The cases used in this study all included an instance or several instances of students making mathematical errors. However, the cases were taken from different sources, such as research articles, the participants' own homework assignments, and classroom observations. In addition, the cases were introduced to participants in different situations such as homework assignments and classroom activities. Studies have shown that one's beliefs and affect may impact on the way that individual engages in professional development (Roesken-Winter, 2013). Thus, if we want to improve prospective teachers' engagement with the cases, it is relevant to investigate their views on the ways in which cases are used in the course. In addition, Ball (1990) claimed that the experiences of prospective teachers during their university methods course may impact on the future teachers' ideas, ways of seeing, and ways of acting. Thus, the way prospective teachers view their experiences with students' errors may impact on the way they see those errors when they become teachers and the way they act on those errors in their future classrooms. Our main research questions are: How do prospective teachers view the impact of cases taken from different sources on their understanding of mathematical errors? How do prospective teachers view the impact of cases presented in different situations on their understanding of mathematical errors?

Using Cases, Events, and Situations in Teacher Education

The idea of using cases in teacher education is not new. Shulman (1986), in his seminal work on teachers' knowledge, argued that although the case method was historically used in teaching law and medicine, it could also be used for teaching future teachers. He used the term "case knowledge" to describe "knowledge of specific, well documented, and richly described events" (p. 11). He warned, however, that a case "is not simply the report of an event or incident. To call something a case is to make a theoretical claim – to argue that it is a 'case of something,' or to argue that it is an instance of a larger class" (p. 11). Furthermore, Shulman claimed that the use of the case method in teacher education can illuminate both the practical and theoretical sides of teaching. In other words, it can help bridge the gap between fieldwork and course work.

Taking into account that not every event may be considered a case, it becomes relevant to discuss how and why certain cases are chosen specifically for the education of future mathematics teachers. Markovits and Smith (2008) describe two kinds of cases used in mathematics teacher education – exemplars and problem situations. Exemplars consist of lengthy descriptions of an entire instructional episode that highlight key ideas about mathematics teaching and learning. Key ideas include not only pedagogical moves but also key mathematical ideas in a specific mathematical domain. For example, cases may be used to demonstrate the crucial role of teachers' actions and their interactions with students during classroom instruction that includes cognitively challenging mathematical tasks (Henningsen & Stein, 1997). Exemplars illustrate authentic practice. They do not necessarily exemplify best practice but may be used in teacher education to study factors which inhibit students' learning (Markovits & Smith, 2008).

The second type of case mentioned by Markowits and Smith (2008) is problem situations. As opposed to the first type described above, problem situations are usually relatively short and may convey real events that took place in a classroom or a hypothetical situation based on research related to students' ways of thinking. In general, they describe classroom events involving mathematics, which raise a problem or dilemma inviting readers to analyze the situation and to suggest ways of responding to the problem (Markovits & Even, 1999).

Different studies described different ways in which cases or situations were used in teacher education. Conner, Wilson, and Kim (2011) developed a tool they called "Situations" which consisted of three parts. The first part contained a prompt which included a brief description of a mathematical event along with students' and teachers' questions and insights. The second part was a description of various aspects of mathematical proficiency relevant to the event. The last part included a commentary which discussed a summary of key ideas to be found in the first two parts. In their study, participants discussed the Situation along with the facilitator, after being given time to individually reflect on the Situation. Pang (2011) used video cases accompanied by comprehensive narratives which included the background of the recorded lesson as well as related theory, when working with prospective teachers. The videos were taken from both planned and unplanned lessons given by experienced teachers and publically available video libraries, as well as recording of the prospective teachers' teaching during their fieldwork. Prospective teachers were required to view recording and read the accompanying text before coming to class and then discuss key elements during the class.

While most studies which investigated the use of cases in teacher education focused on promoting teachers' mathematical and pedagogical content knowledge, some studies also noted affective issues. Conner et al. (2011) reported that prospective teachers said that engagement with Situations was one of the most helpful aspects of their methods course. Furthermore, the instructor and prospective teachers found the Situations to be very relevant. While discussing video cases, Santagata and Guarino (2011) mentioned that when participants view exemplar cases, there can be a problem with the "distance PSTs [preservice teachers] might feel between their teaching abilities and the ability of the teachers portrayed in the videos" (p. 143). Lin (2005) remarked that the "video-cases motivated preservice teachers to rethink the importance of a student-oriented approach and to emphasize the need for engaging students with challenging mathematical tasks" (p. 372). Working with practicing teachers, Walen and Williams (2000) suggested that discussions and reflections of cases elicited powerful reactions among the teachers and that the cases played a surprisingly powerful role in helping the teachers acknowledge and deal with their classroom concerns. In other words, using cases in teacher education has the potential to affect participants' engagement in their learning as well as their beliefs and practice regarding teaching mathematics.

Methodology

The Teacher Preparation Program and Course

Participants in this study were 31 students enrolled in a university program for preparing secondary school mathematics teachers. All students had a first degree in mathematics or a mathematically rich field such as engineering and after successfully completing the program would receive a teaching license. The program included attending university courses as well as doing 130 h of fieldwork in secondary school mathematics classes under the guidance of a mathematics teacher. In addition, all participants attended a workshop at the university, run by an expert mathematics teacher, who discussed with the participants their fieldwork experience. The expert teacher at the university was in contact with other university lecturers who taught these participants. In general, the aims of the university courses were to promote participants' mathematics content knowledge and pedagogical knowledge for teaching (Ball et al., 2008).

This study focuses on a semester-long university course, given by a senior university lecturer, which emphasized the analysis of students' mistakes as a way for promoting prospective teachers' knowledge of common student errors and, in addition, as a way for promoting participants' mathematics knowledge. During the course, which met once a week for 2 h, students were introduced to theories for analyzing the reasons behind students' common mistakes such as the conflict between concept images and concept definitions (Tall & Vinner, 1981), intuitive rules (Stavy & Tirosh, 2000), and the interaction between formal, algorithmic, and intuitive elements of mathematics (Fischbein, 1993).

The mathematical errors analyzed during the course came from different sources and were presented and discussed among participants in different ways. In total, there were six different error analysis situations. The sources of the first two situations were errors made by the participants of the course. Every week, students were given a series of four to five mathematics problems to solve at home, and then during the class, the lecturer went over the participants' correct and incorrect solutions. Thus the source of the first situation was the participants' homework errors. At times, a new problem was presented in class, and students solved the problem in class, with some participants solving the problem on the whiteboard up front. Thus, the second source was participants' errors made during the class. The source of the third and fourth error situations came from research papers. From the fourth lesson on, two students presented to the class their summary of common mistakes taken from a research paper assigned to them by the lecturer. The third source came from research papers describing quantitative studies of students' common errors. The fourth source came from papers describing classroom interactions and a qualitative analysis of a case which involved students' mathematical errors. The fifth and sixth sources came from the participants' current experiences with secondary school students. During the semester, participants took part in classroom observations and were required twice during the semester to report and analyze cases taken from those observations. The fifth error situation was a case chosen by the lecturer to be used in a homework assignment. All participants were required to analyze errors which arose during the case, according to theories learned during the course. The sixth source came from the participants' experience with students. Participants were given two mathematics problems known to cause student errors. Participants were required to ask two high school students to solve those problems, interview the students, and then analyze the students' solutions, including errors that arose during the solution process.

Research Tool

At the end of the course, students were requested to fill out a questionnaire that had the following instructions:

> During the course, you had the opportunity to analyse incorrect solutions which arose in different situations. For each situation, rate the extent to which the activity helped you to understand error analysis: greatly, to some extent, a little bit, not at all.

Following this instruction was a list of the six different situations, as described above. The lecturer also clarified each situation orally ensuring that all of the participants recognized the different situations. At the end of the six situations, participants were asked to answer the following question: If you had to choose only two situations (from the above six) which would you choose?

Results

Recall that participants were requested to rate the extent of each of the six situations' impact on their understanding of common mathematical errors. Each rating was assigned a numerical value: 1, not at all; 2, a little bit; 3, to some extent; and 4, greatly. To begin with, we note that only one participant rated one situation (Situation 6) as not having any impact at all. In other words, results indicated that prospective teachers perceive the use of cases, regardless of the source of the case or the

Table 1 Mean ratings of each situation's impact ($N = 31$)

Source of mistakes	Mistakes made by participants		Mistakes reported in research papers		Eyewitness to others' mistakes	
Situation	1 H.W.	2 Classwork	3 Quantitative study	4 Qualitative study	5 Classroom observation	6 Student interview
Mean	3.84	3.52	3.39	3.13	3.39	2.97
SD	0.37	0.57	0.72	0.72	0.72	0.84

situation in which it is presented, as having an impact on their understanding of mathematical errors. Furthermore, the mean ratings (see Table 1) indicate that overall, the use of cases was perceived as being more than just a little bit meaningful.

Taking a closer look at the results, we note slight differences. Situation 1, where mistakes made by participants in their homework assignments were discussed in class, was perceived as having the most impact on participants' understanding of the errors, followed by mistakes made by participants while engaged in classwork. In other words, participants viewed that analyzing their own mathematical mistakes was more meaningful than analyzing mistakes made by others. Participants viewed Situations 3 and 5 as having less of an impact than Situations 1 and 2 but more than Situations 4 and 6.

That participants viewed Situations 3 and 5 as having the same impact on their learning is surprising. First, the errors presented in the two situations come from very different sources. Situation 3 consisted of cases taken from research papers that reported on quantitative studies of students' common mistakes, while Situation 5 was one case taken from a participant's classroom observation. Second, the way each situation was used was also different. Situation 3 was discussed in class, and Situation 5 was a homework assignment. While the same surprise might be felt for the similar preference for Situations 4 and 6, those situations are at least similar in that both deal in depth with one or two students and the mathematical errors those few students made. In essence, it may be said that Situation 4 prepared them to deal with Situation 6 although Situation 6 was more personal in that the participant actually conducted firsthand an informal qualitative study.

Although participants were not requested to explain their ratings, some participants did add clarifications. For example, regarding the high rating for the first two situations, one participant wrote for Situation 1, "the attention given to our solutions sharpened my understanding of mistakes and corresponding theories." Regarding Situation 2 that same participant wrote, "when we talk about 'our' solutions, I relate better to the material." One participant commented on his or her high rating for Situation 6, "when I interviewed the students, I went on to other topics which sharpened my understanding of the source of those mistakes and this will help me in my teaching." In other words, this participant used the interview situation as an opportunity to test out other theories and thereby strengthen knowledge gained regarding errors.

For the most part, as noted above, Situation 6 received relatively low ratings. This may be explained by one participant's comment: "We only had to interview

Table 2 Frequency (%) of participants' choices for most preferred situations ($N = 31$)

Source of mistakes	Mistakes made by participants		Mistakes reported in research papers		Eyewitness to others' mistakes	
Situation	1 H.W.	2 Classwork	3 Quantitative study	4 Qualitative study	5 Classroom observation	6 Student interview
Frequency	25 (81)	11 (35)	12 (39)	4 (13)	6 (19)	2 (6)

two people. Perhaps if we interviewed more people it would have helped more in our understanding of students' common mistakes (two is not a representative sample)." While this comment was given by only one participant, it hints at a possible reason for the relatively low ratings given to both Situations 4 and 6. In their attempt to understand why some errors are more common than others, perhaps participants may have felt the need to read about or to experience many students making the same error as opposed to hearing about or even personally interacting with one or two who made those errors. On the other hand, another participant wrote that the reports on quantitative studies were less effective than the reports from qualitative studies because, "in the case of the quantitative studies, we did not go into depth, and not enough time was given, but for the other papers we analysed concrete and clear mistakes and that helped our understanding." It is not clear from this comment if the participant means that the quantitative studies dealt with too many errors at once and thus it was impossible to analyze and understand all of them in depth or, for some reason, during the class, there was less time devoted to those papers. In any case, for this participant, it was important to understand each mistake in detail.

On the second part of the questionnaire, participants were asked to choose the two situations they prefer most. Two participants chose only one situation. Table 2 presents the number of teachers who chose each situation as one of their most preferred. As can be seen from Table 2, Situation 1 was chosen by over three-quarters of the teachers, Situations 2 and 3 by approximately a third of the participants, and the rest by even less participants. Combining situations from the same source, 27 participants (87%) would choose at least one case stemming from their own mistakes (Situations 1 or 2), 15 (48%) participants would choose at least one situation based on errors reported in a research paper (Situations 3 and 4), and 8 (26%) participants would choose cases based on others' mistakes that they or their peers had witnessed.

In light of the responses to the first part of the questionnaire, it is not surprising that so many participants chose the first situation as one of the two they most preferred. One participant who chose Situation 1 commented:

> Sometimes, when working on the assignments, we discussed problems that were more complex than the usual broad common mistakes and you could get lost. Still, the solutions and their analysis were a tool that allowed me to solidify my understanding of the concept and the mistake by going over it several times.

That same participant also chose Situation 3 and wrote:

> Relating to common mistakes found in research allowed me to focus on the specific mathematical mistake (and not on all kinds of different mistakes) that stem from a specific mathematical problem. For example, the intuitive rule 'More of A, More of B', when talking about a simple function such $f(x) = 0.5^x$ and the question of which is greater $f(2)$ or $f(1)$. Together with the detailed frequencies and the discussion of the students' explanations (along with the researcher's analysis), helped me to understand better this type of error.

Two differences between the results of the first and second parts of the questionnaires can be seen. First, on the second part of the questionnaire, only a third of the participants chose Situation 2 as one of the two most preferred situations, while almost all of the participants gave this situation a high impact rating on the first part of the questionnaire. It could be that participants viewed the first two situations as being very similar, and thus if they could only choose two situations and they already chose Situation 1, then there was no need to also choose Situation 2. The second difference between the two parts of the questionnaires was participants' responses to Situations 3 and 5. While on the first part of the questionnaire, participants gave these situations similar impact ratings, on the second part, twice as many chose Situation 3 as Situation 5.

Summary, Discussion, and Implications

The first question addressed by this paper was: How do prospective teachers view the impact of cases taken from different sources on their understanding of mathematical errors? The errors in the cases came from three sources: the participants' own errors (Situations 1 and 2), errors reported in research studies (Situations 3 and 4), and participants' observations of others' errors (Situations 5 and 6). Findings from both parts of the questionnaire indicated that participants' viewed learning from their own errors as most meaningful. When it came to choosing only two situations, most participants chose Situation 1 (learning from mistakes they made when solving homework assignments) over Situation 2 (learning from mistakes made during classwork). Although we cannot know for sure the reasons for this preference, it could be that there was more anonymity in discussing mistakes made in the privacy of one's home than mistakes made in class. It could also be that participants had more time at home to work on mathematics problems, and thus discussing those problems was more meaningful than discussing mistakes made on the spot during class. Finally, mistakes taken from the homework assignments were specifically chosen by the lecturer for discussion because at least several participants made the same mistake and thus prospective teachers could relate to those mistakes.

Regarding learning from mistakes made by others, on the one hand, participants found that learning about mistakes from research papers and learning from observations were both meaningful. On the other hand, if they had to choose one or the other, most would choose cases based on errors reported in research studies. This last result is surprising because learning from research studies was thought to be more connected with theory and rather removed from practice, while analyzing errors made by students whom the participants directly interviewed was thought to bring practice closer to the theory discussed in the course. According to Zeichner (2010), bridging the gap between theory and practice is important and is thought to impact greatly on teacher preparation. In trying to understanding this result, we take a look at the differences between the ways these cases were presented and used in the course and now turn toward the second research question.

The second question addressed by this paper was: How do prospective teachers view the impact of cases presented in different situations on their understanding of mathematical errors? We begin with the problem of why participants preferred cases taken from research papers over cases based on their own observations. The cases based on research papers were presented by different participants to the whole class and then discussed and analyzed in the course along with the teacher educator and the other participants. The cases based on the participants' own observations were analyzed by individual participants as part of a homework assignment and as part of the final project. Although the participants received feedback on their work, these cases were not discussed by the whole group of participants along with the teacher educator. While researchers agree that it is important for prospective teachers to observe and reflect on students' mathematical thinking (e.g., Ball et al., 2008), one of the problems with fieldwork placements is that the prospective teachers lack a common experience to discuss with their peers (Masingila & Doerr, 2002). The same might be said for analyzing students' mathematical thinking through the analysis of cases. Students preferred to learn from situations that were discussed together in class (Situations 1, 2, 3, and 4) over situations that were analyzed alone (Situations 5 and 6).

There are several possible implications of this study for teacher education. First, the use of cases, in general, is seen as a positive learning activity for prospective teachers. However, not all cases have the same impact. Although participants agreed that it was fruitful to learn from their own mistakes, they preferred to learn from mistakes they made individually but then discussed collectively. Thus, teacher educators might take into consideration prospective teachers' comfort zone when discussing mistakes in class as well as their need to discuss mistakes with their peers. Second, although bringing practice closer to theory is important, it does not preclude learning from previous research reports. This is in line with Tsamir (2008) who showed the effectiveness of using theories as tools in teacher education. Finally, that participants felt it was less meaningful to analyze errors made by students they interviewed may inform teacher educators who work with prospective teachers that have limited access to field practice.

Studying cases can support prospective teachers' analysis and reflection of their own emerging practices (Masingila & Doerr, 2002). Thus, teachers' preferences for the different case sources and different ways of working with the cases may impact on the way participants will use students' mistakes during their future practice. The positive views prospective teachers had with regard to learning from their own mistakes may encourage these participants to build on their future students' mistakes as part of their future teaching practice, instead of trying to avoid or simply "fix" mistakes. Kaur (2009) suggested that an important element of good teaching practice is to encourage students to learn from their mistakes, not only by stressing the final answer, but by focusing on the kinds of mistakes made. Similarly, participants added comments on their questionnaires stressing the importance of analyzing the errors and not just fixing them. These positive experiences will hopefully impact on their future teaching.

Acknowledgment This research was supported by the Trump Foundation grant no. 145.

References

Ball, D. L. (1990). Breaking with experience in learning to teach mathematics: The role of a pre-service methods course. *For the learning of mathematics, 10*(2), 10–16.

Ball, D. L., & Forzani, F. M. (2009). The work of teaching and the challenge for teacher education. *Journal of Teacher Education, 60*(5), 497–511. https://doi.org/10.1177/0022487109348479

Ball, D. L., Thames, M. H., & Phelps, G. (2008). Content knowledge for teaching: What makes it special? *Journal of Teacher Education, 59*(5), 389–407. https://doi.org/10.1177/0022487108324554

Conner, A., Wilson, P. S., & Kim, H. J. (2011). Building on mathematical events in the classroom. *ZDM, 43*(6–7), 979–992. https://doi.org/10.1007/s11858-011-0362-1

Fischbein, E. (1993). The interaction between the formal, the algorithmic and the intuitive components in a mathematical activity. In R. Biehler, R. Scholz, R. Strässer, & B. Winkelmann (Eds.), *Didactics of mathematics as a scientific discipline* (pp. 231–245). Dordrecht, The Netherlands: Kluwer.

Henningsen, M., & Stein, M. K. (1997). Mathematical tasks and student cognition: Classroom-based factors that support and inhibit high-level mathematical thinking and reasoning. *Journal for Research in Mathematics Education, 28*, 524–549. https://doi.org/10.2307/749690

Kaur, B. (2009). Characteristics of good mathematics teaching in Singapore grade 8 classrooms: A juxtaposition of teachers' practice and students' perception. *ZDM, 41*(3), 333–347. https://doi.org/10.1007/s11858-009-0170-z

Lin, P. J. (2005). Using research-based video-cases to help pre-service primary teachers conceptualize a contemporary view of mathematics teaching. *International Journal of Science and Mathematics Education, 3*(3), 351–377. https://doi.org/10.1007/s10763-004-8369-5

Markovits, Z., & Even, R. (1999). The decimal point situation: A close look at the use of mathematics-classroom-situations in teacher education. *Teaching and Teacher Education, 15*(6), 653–665. doi.org/10.1016/S0742-051X(99)00020-7

Markovitz, Z., & Smith, M. (2008). Cases as tools in mathematics teacher education. In D. Tirosh & T. Wood (Eds.), *The international handbook of mathematics teacher education* (Vol. 2, pp. 39–64). Rotterdam, The Netherlands: Sense Publishers.

Masingila, J. O., & Doerr, H. M. (2002). Understanding pre-service teachers' emerging practices through their analyses of a multimedia case study of practice. *Journal of Mathematics Teacher Education, 5*(3), 235–263. https://doi.org/10.1023/A:1019847825912

Pang, J. (2011). Case-based pedagogy for prospective teachers to learn how to teach elementary mathematics in Korea. *ZDM, 43*(6–7), 777–789. https://doi.org/10.1007/s11858-011-0352-3

Roesken-Winter, B. (2013). Capturing mathematics teachers' professional development in terms of beliefs. In *Proficiency and beliefs in learning and teaching mathematics* (pp. 157–178), Rotterdam, The Netherlands: Sense Publishers. https://doi.org/10.1007/978-94-6209-299-0_11

Santagata, R., & Guarino, J. (2011). Using video to teach future teachers to learn from teaching. *ZDM, 43*(1), 133–145. https://doi.org/10.1007/s11858-010-0292-3

Shulman, L. S. (1986). Those who understand: Knowledge growth in teaching. *Educational Researcher, 15*(2), 4–14.

Stavy, R., & Tirosh, D. (2000). *How students (mis-) understand science and mathematics: Intuitive rules*. Teachers College Press, NY, USA.

Tall, D., & Vinner, S. (1981). Concept image and concept definition in mathematics with particular reference to limits and continuity. *Educational Studies in Mathematics, 12*(2), 151–169. https://doi.org/10.1007/BF00305619

Tsamir, P. (2008). Using theories as tools in mathematics teacher education. In D. Tirosh & T. Wood (Eds.), *The international handbook of mathematics teacher education* (Vol. 2, pp. 211–234). Rotterdam, The Netherlands: Sense Publishers.

Walen, S. B., & Williams, S. R. (2000). Validating classroom issues: Case method in support of teacher change. *Journal of Mathematics Teacher Education, 3*(1), 3–26. https://doi.org/10.1023/A:100991751031

Zeichner, K. (2010). Rethinking the connections between campus courses and field experiences in college-and university-based teacher education. *Journal of Teacher Education, 61*(1–2), 89–99. https://doi.org/10.1177/0022487109347671

Developing and Trialling a Simple-to-Use Instrument for Surveying Teacher Education Students' Mathematics-Related Beliefs

Judy Sayers and Paul Andrews

Abstract Acknowledging that what teachers believe informs how they teach, we argue the importance of understanding, at both entry and exit, teacher education students' beliefs about mathematics, mathematics teaching and themselves as learners of mathematics. In this paper we report on the development and trial of a simple-to-use online survey instrument focused on uncovering teacher education students' mathematics-related beliefs. Twenty items, targeted on a range of constructs reported in the literature, were set against five-point Likert scales. The instrument was found to be reliable, and an exploratory factor analysis yielded seven interpretable belief dimensions. The interactions of these dimensions allude to groups of students likely to prove problematic during their programme.

Keywords Factor analysis • Mathematics-related beliefs • Preservice teacher education • Primary mathematics • Survey research • Sweden

Introduction

In this paper we explore the mathematics-related beliefs of Swedish primary teacher education students (Teeds). Our aims were essentially twofold. Firstly, teachers' beliefs determine what and how they teach (Wilson & Cooney, 2002). Therefore, eliciting Teeds' mathematics-related beliefs should facilitate the identification of students with undesirable or potentially problematic beliefs. Secondly, as in many other developed countries, many Teeds fail to complete their programmes (Stokking, Leenders, de Jong, & van Tartwijk, 2003). A better understanding of student beliefs should help identify Teeds at risk of failing to complete their course. In the following we report on the development and trial of a simple-to-use instrument for

J. Sayers (✉) • P. Andrews
Stockholm University, Stockholm, Sweden
e-mail: judy.sayers@mnd.su.se

H. Palmér, J. Skott (eds.), *Students' and Teachers' Values, Attitudes, Feelings and Beliefs in Mathematics Classrooms*,
https://doi.org/10.1007/978-3-319-70244-5_8

evaluating a range of mathematics-related beliefs known to have an impact on later teaching.

Mathematics-Related Beliefs

Research into mathematics-related beliefs occupies a large field in the countryside of mathematics education research. Much of this has addressed school students because their experientially derived beliefs underpin their reactions to the opportunities they receive (Callejo & Vila 2009). In general, belief-related research focuses on either belief as cognitive or belief as affective. In terms of the cognitive, research has explored several related fields. For example, some have examined students' epistemological beliefs "about the source, certainty, and organization of knowledge" (Chan & Elliott, 2002, p.393). Others have studied students' efficacy beliefs or the "beliefs in one's capabilities to organize and execute courses of action required to manage prospective situations" (Bandura, 1997, p.2). Still others have explored students' motivation beliefs tied to whether they are driven by mastery or performance goals (Gordon, Dembo, & Hocevar, 2007) or beliefs about the role of the teacher in the construction of learning (Lloyd, 2005). With respect to affect, researchers tend to focus on the emotions that typically derive from students' experiences of and inform future responses to mathematics (Op't Eynde, De Corte, & Verschaffel, 2002). These automated emotional responses are varied and may derive from teachers' tones of voice or repeated failure on particular forms of task (Evans, Morgan, & Tsatsaroni, 2006). Thus, emotions underpin one's motivation (Hannula, 2006).

With respect to Teeds' mathematics-related beliefs, research has been driven by the argument that a better understanding of how beliefs and subsequent practice interact should facilitate better-focused teacher education programmes (Cooney, Shealy, & Arvold, 1998). Frequently, Teeds' beliefs match those of school students. For example, many primary Teeds begin their courses with high levels of anxiety towards and negative perception of mathematics (Ambrose, 2004; Hembree, 1990). They lack a belief in its importance, especially with respect to younger children, seeing mathematics as an abstract subject requiring rote memorisation of procedures and formulae (Hannula, 2002). These problems are compounded by a fear of failure that undermines Teeds' personal learning of mathematics and leads to inadequate teaching (Hannula, 2002; Hembree, 1990).

Much research on Teeds' epistemological beliefs has drawn on the four factors identified in Schommer's (1990) American study. These are the *ability to learn is innate*, *knowledge is discrete and unambiguous*, *learning is quick or not at all* and *knowledge is certain*. For example, Cheng, Chan, Tang, and Cheng (2009) evaluated Hong Kong Teeds' epistemological and teaching-related beliefs, while Aypay (2010) repeated the same process in Turkey. Both studies found that Teeds who view knowledge as fixed tend to have traditional beliefs about teaching, while those who saw knowledge as malleable tend to have constructivist beliefs.

Framing Our Conceptualization of Beliefs

In defining what we understand a belief to be, we are conscious of concerns that research has been located in definitional inconsistency (Pajares, 1992). Typically, the debate has focused on distinctions between belief and knowledge, although some have suggested a hierarchy rather than a distinction in that experientially formed beliefs incorporate knowledge even though the latter, which requires consensus, holds a higher epistemological warrant than the former (Op't Eynde et al., 2002). Other distinctions have been proposed between testable beliefs derived from immediate experience of an object and distal beliefs based on a remote experience of the same object (Abelson, 1979) and conscious and unconscious beliefs (Ernest, 1989). Others have expressed concern over the evaluative nature, impermanence and function of beliefs, as well as the unclear relationship between beliefs, attitudes and values (Hannula, 2006; Pajares, 1992). In this study we argue for an inclusive definition. That is, beliefs are "understandings, premises, or propositions about the world that are felt to be true" (Richardson 1996, p.103). With respect to mathematics, we accept that beliefs are "the implicitly or explicitly held subjective conceptions students hold to be true that influence their mathematical learning and problem-solving" (Op't Eynde et al., 2002, p.16). Such definitions avoid many of the concerns above and acknowledge Green's (1971) argument that beliefs are not dichotomous but lie on a continuum from the unconscious centre to the conscious periphery.

Methods

As indicated, the purpose of this paper is to report on the development of a simple-to-use instrument for evaluating Teeds' mathematics education-related beliefs. Drawing on evidence that particular forms of belief are more likely to influence practice, we drew on extant instruments (Andrews & Diego-Mantecón, 2015; Chan & Eliott, 2004; Op't Eynde et al., 2002) to develop a 20-item instrument focused on beliefs about the nature of mathematics, oneself as a learner of mathematics, mathematics teaching and so on. Each item was placed against a five-point scale from strongly agree to strongly disagree.

Students, all studying to be specialist teachers of mathematics, were invited to complete the survey online during the introductory sessions of their mathematics education courses. Students were asked to include their names and university registration numbers to facilitate a repetition of the survey at the end of their course. As a consequence, students were promised that they would not be mentioned in any report. Participation was voluntary, and 95% of students aiming to teach the lower primary grades (F-3) completed the survey, as did 76% of students preparing to teach the upper primary grades (4–6). This paper is based on data from four cohorts, two for each programme beginning in 2014 and 2015. There were 174 students following the lower primary route and 143 the upper secondary route, giving a total of 317 students, which is sufficient for a factor analysis based on a survey of 20 items (Field, 2005).

Data were imported into SPSS® and a principal component with varimax rotation analysis undertaken to determine "which belief categories and subcategories have empirical grounds" (Op't Eynde et al., 2006, p.65). This decision was based on the fact that all the scales included in the pilot comprised a reduced number of items from their originals and that none had ever been used with Swedish students, making confirmatory factor analyses inappropriate. Finally, we adopted $p < 0.01$ as the threshold for rejecting null hypotheses, facilitating our reporting of practical and not just statistically significant differences.

Results

Drawing on conventional practices of extracting factors with eigenvalues greater than one, meaning that each factor accounted for more variance than that of a single item, a seven-factor solution was obtained that accounted for 60.2% of the total variance. In identifying factors, we erred on the side of caution and accepted only those items with a factor loading greater than or equal to 0.4. Details of this can be seen in Table 1; four items, which we discuss below, were eliminated by the factor analysis. All factors were straightforwardly interpretable and labelled, with acronyms, as in Table 2.

Refining the Instrument

The four eliminated items were *school mathematics is taught so that students get a good job; children learn mathematics best by listening to a clear explanation from the teacher; when solving mathematical problems, I try to identify the similarities with or differences from problems I have solved before;* and *when I get stuck with a mathematical problem, I keep trying to think of different ways without giving up.* Each of these items either failed to load on any factor, but came close to loading on two or three, or loaded on at least two. In short, students' responses to all four indicated some sense of ambiguity. For example, was the third item focused on identifying similarities with earlier problems or differences between earlier problems? Similarly, was the fourth item focused on different problem-solving strategies or personal persistence? In short, the analyses confirmed that ambiguous items have no place in such instruments.

Table 1 Items that had factor loadings

	MS	AT	LT	FP	IM	IA	DL
I was good at school mathematics	.838						
I liked studying school mathematics	.805						
I worked well in mathematics lessons	.752						
Children learn mathematics best when they read through worked examples in a textbook		.791					
Children learn mathematics best by working through questions on their own		.636					
Children learn mathematics best when they ask the teacher for help in lessons		.614					
School mathematics is taught to develop the ability to think logically			.791				
School mathematics is taught because it will be useful in everyday life			.786				
When I have solved a mathematical problem, I look for different ways to solve it				.790			
When I am unable to solve a mathematical problem, I reflect on why I did not solve it				.753			
I want to study mathematics so that I can be a good teacher of mathematics					.770		
It is important to study school mathematics					.736		
Ability in mathematics is something that you either have or you haven't						.729	
It is possible to improve in mathematics by working hard						−.530	
Children learn mathematics best by discussing problems with their friends							.828
Children learn mathematics best when they have to explain their thinking to a friend							.787

Table 2 Factor descriptions and acronyms

Factor description	Acronym
I was mathematically successful at school	MS
Children learn best when working independently	IL
School mathematics provides an important life tool	LT
I am a flexible problem solver	FP
It is important to study school mathematics	IM
Mathematical success requires innate ability	IA
Mathematics learning should be discursive	DL

General Results

The seven factors represent a range of belief-related constructs. For example, *I was mathematically successful at schoo*l (MS) and *I am a flexible problem solver* (FP) represent different perspectives on mathematical self-concept. Two others, school mathematics provides an important life tool (LT) and *it is important to study school mathematics* (IM), offer perspectives on the purpose of school mathematics, one general and the other particular. Two more, *children learn best when working independently* (IL) and *mathematics learning should be discursive* (DL), offer different didactical perspectives. Finally, *mathematics success requires innate ability* (IA) offers a perspective on the nature of intelligence. In short, the seven factors offer perspectives on at least four distinct forms of belief function.

The figures of Table 3 show the mean score for all students on each of the seven factors. Acknowledging that scores fall in a range from one to five, with three representing some sense of neutrality, it can be seen that on all factors but one, *mathematical ability requires innate ability*, students scored positively. For example, their scores on LT and IM indicate that they have similarly strong views with respect to the purpose of school mathematics. From a didactial perspective, their scores on IL and DL are also positive. Moreover, their scores on MS and FP imply that students are generally positive with respect to their personal mathematical competence.

Particular Results

However, if such surveys are to be meaningful, particularly with respect to identifying Teeds in need of support, then it is important to identify subgroups with beliefs unsupportive of the development of equal opportunities to mathematics learning environments. To this end the sample was split on each factor according to whether a student scored at the hypothetical neutral or above (factor scores greater than or equal to 3) or below the neutral (factor scores less than 3). Scores for each other factor were then calculated according to this split and can be seen in Tables 4, 5, 6 and 7, respectively. The presentation of the results focuses on the negatively scoring group in each case.

Table 3 Mean scores for the seven factors

	Mean	s.d.
MS	3.45	0.90
IL	3.27	0.68
LT	4.05	0.80
FP	3.32	0.81
IM	4.48	0.52
IA	2.43	0.69
DL	3.89	0.68

Table 4 Factor scores for students in split groups according to MS and IL

Split by MS (−84 and +233)					Split by IL (−81 and +236)				
	Mean	Mean	*t*	*p*		Mean	Mean	*t*	*p*
MS	2.22	3.89	27.89	0.000	MS	3.29	3.50	1.80	0.073
IL	3.07	3.34	3.17	0.002	IL	2.38	3.58	25.62	0.000
LT	3.80	4.13	3.00	0.003	LT	3.98	4.07	0.84	0.401
FP	2.71	3.54	8.94	0.000	FP	3.18	3.37	1.84	0.067
IM	4.42	4.51	1.26	0.207	IM	4.50	4.48	0.32	0.753
IA	2.40	2.44	0.43	0.664	IA	2.02	2.57	7.36	0.000
DL	3.77	3.93	1.87	0.063	DL	3.83	3.91	0.89	0.372

Table 5 Factor scores for students in split groups according to LT and FP

Split by LT (−22 and +295)					Split by FP (−95 and +222)				
	Mean	Mean	*t*	*p*		Mean	Mean	*t*	*p*
MS	2.84	3.49	3.30	0.001	MS	2.83	3.71	8.77	0.000
IL	3.13	3.28	1.06	0.290	IL	3.07	3.36	3.48	0.001
LT	2.07	4.19	30.69	0.000	LT	3.84	4.13	3.01	0.003
FP	3.02	3.34	1.83	0.068	FP	2.35	3.73	25.76	0.000
IM	4.07	4.52	3.97	0.000	IM	4.42	4.51	1.53	0.127
IA	2.44	2.43	0.07	0.942	IA	2.37	2.45	1.02	0.310
DL	3.66	3.91	1.35	0.192	DL	3.84	3.91	0.85	0.398

Table 4 shows the results of splitting the sample on MS and IL. With respect to their being *mathematically successful*, it can be seen that 84 students (26.5%) saw themselves as mathematically unsuccessful. Moreover, these MS-negative students not only saw themselves as weak on FP but scored significantly lower on IL and LT than their more mathematically successful colleagues. On other factors, their scores were lower but not significantly different in every case. With respect to *mathematics as best learnt independently*, 81 students (25.5%) scored negatively. Moreover, these LT-negative students scored significantly more negatively than their peers with respect to IA. That is, a negative belief with respect to mathematics as *best learnt independently* was matched by a strong negative belief that *mathematical ability is innate*. Finally, while not significant, all other factors achieved lower scores than their positively scoring colleagues.

Table 5 shows the results of splitting the data on LT and FP. With respect to *mathematics as a life tool*, only 22 (7%) students scored negatively, a view implying a limited understanding of how school mathematics can support learners in their later lives. These LT-negative students also scored negatively with respect to MS and IA, although the latter was effectively identical to the score achieved by LT-positive students. These same LT-negative students responded significantly less positively with respect to IM than their colleagues. On all other factors, these students scored lower than their positive colleagues. With respect to being flexible problem solvers, 95 students (30%) scored negatively. Moreover, reciprocating the results of Table 4,

Table 6 Factor scores for students in split groups according to IM and IA

Split by IM (−3 and +314)					Split by IA (−229 and +88)				
	Mean	Mean	*t*	*p*		Mean	Mean	*t*	*p*
MS	3.08	3.45	0.70	0.486	MS	3.45	3.43	0.13	0.894
IL	2.83	3.28	1.13	0.260	IL	3.13	3.64	6.97	0.000
LT	3.83	4.05	0.46	0.643	LT	4.07	3.98	0.95	0.344
FP	3.00	3.32	0.69	0.493	FP	3.30	3.38	0.82	0.412
IM	2.33	4.50	7.83	0.000	IM	4.52	4.39	1.96	0.051
IA	2.44	2.43	0.04	0.969	IA	2.09	3.32	26.82	0.000
DL	3.00	3.90	2.31	0.021	DL	3.87	3.95	1.04	0.301

Table 7 Factor scores for students in split groups according to DL

Split by DL (−18 and +299)				
	Mean	Mean	*t*	*p*
MS	3.18	3.46	1.28	0.201
IL	3.06	3.29	1.40	0.162
LT	3.64	4.07	2.25	0.025
FP	3.00	3.34	1.73	0.085
IM	4.36	4.49	0.70	0.493
IA	2.54	2.42	0.68	0.497
DL	2.17	3.99	14.31	0.000

these FP-negative students also scored significantly negatively with respect to MS. In addition, they were significantly less positive than their FP-positive colleagues with respect to IL and LT. Finally, they scored lower than their colleagues on the remaining three factors.

Table 6 shows the results of splitting the data on IM and IA. In terms of *the importance of school mathematics*, only three students scored negatively. Moreover, while the number of students involved was too small for any differences to be significant, these students' scores on all other factors were lower than their IM-positive peers with the exception of IA, which was effectively identical. With respect to *mathematical ability being innate,* 229 students scored negatively, leaving 88 students with a positive score. That is, 27.8% of the cohort seem to have a belief in intelligence as fixed. These IA-positive students had a significantly more positive score on IL, indicating a belief that mathematics should be learnt independently. On the remaining factors, this group's scores were neither significantly nor practically different from their IA-negative colleagues. Finally, Table 7 shows the results of splitting the data on *mathematics is best learnt discursively.* Here, only 18 students (5.7%) scored negatively, and, with the single expection relating to IA, this group scored lower, but not significantly so, on all other factors than their DL-positive colleagues.

Discussion

In this paper we have described the derivation and evaluation of a simple-to-use instrument for analyzing different dimensions of beginning primary teachers' mathematics education-related beliefs. The principal components analysis identified seven factors that not only proved simply interpretable but also yielded different mathematics education-related belief constructs. For example, two factors, *I was mathematically successful at school* and *I am a flexible problem solver*, drew on Teeds' perceptions of their mathematical competence and can be construed as simple measures of mathematics self-concept (Trautwein, Lüdtke, Köller, & Baumert, 2006). Perhaps unsurprisingly these had a positive reciprocal relationship. However, a high number of Teeds scored negatively on each factor (84 and 95 students respectively), confirming that many primary Teeds begin their courses with negative perception of themselves with regard to mathematics (Ambrose, 2004).

Two other factors, *children learn best when working alone from the textbook* and *mathematics learning should be discursive*, focused on beliefs pertaining to how students learn. These also seemed to have a positive reciprocal relationship, which is interesting in that the former reflects traditional perspectives on teaching and learning, while the latter resonates with reform (Chan & Elliott, 2004). Admittedly, while one might argue that the dichotomisation of educational practices is unhelpful, it does seem incongruous that many Teeds believe that learning is achieved through independent and collaborative work simultaneously. Such beliefs confirm that beliefs form clusters, each focused on a particular phenomenon, that allow an individual to hold apparently conflicting beliefs (Abelson, 1979). However, 18 students viewed DL negatively, and while this is proportionately a small group (6%), it indicates that every teacher education class is likely to have one Teed who either lacks an awareness of or does not believe in the role of discussion in children's learning. Of equal concern is the evidence that 236 students (74%) believe mathematics is best learnt through independent working, beliefs likely to be born of their experiences of the typical Swedish mathematics lesson in which a teacher "goes through" something before students work independently from their book (Nosrati & Andrews, 2016).

Two factors, *school mathematics provides an important lifetool* and *it is important to study school mathematics*, offered insights into Teeds' beliefs about the purpose of school mathematics. In both cases, the vast majority of Teeds presented positive perspectives indicative of a general agreement, albeit unspecified, that mathematics is important in the preparation of children for a life beyond school, views commonly held by teachers in some countries but not others (Andrews, 2007). With respect to the former, 22 students (7%) viewed LT negatively, confirming that, as indicated above with respect to DL, each teacher education class will have at least one sceptic with respect to the role of mathematics in children's later lives. The number of students (3) who viewed IM negatively was too small to warrant further comment.

The seventh factor, *mathematical success requires innate ability*, exposed Teeds' beliefs on intelligence. In this regard, 88 students (28%) indicated positively that intelligence is fixed and unchangeable. Such beliefs, resonant with an earlier study of inexperienced Swedish teachers (Jonsson, Beach, Korp & Erlandson, 2012), lead teachers to comfort rather than challenge their pupils, have the effect of demotivating children (Rattan, Good, & Dweck, 2012). In sum, the form of these seven factors and the manner of their interactions showed, at this early stage of these Teeds' professional development, beliefs about the teaching and learning of mathematics in need of challenge. Thus, our objectives seem to have been satisfied; the instrument is simple to use and effective. In the next phase of our work, we will repeat the survey at the end Teeds' programme to determine what changes, if any, have occurred.

References

Abelson, R. (1979). Differences between belief systems and knowledge systems. *Cognitive Science, 3*(4), 355–366. https://doi.org/10.1207/s15516709cog0304_4

Ambrose, R. (2004). Initiating change in prospective elementary school teachers' orientations to mathematics teaching by building on beliefs. *Journal of Mathematics Teacher Education, 7*(2), 91–119. https://doi.org/10.1023/B:JMTE. 0000021879.74957.63

Andrews, P. (2007). The curricular importance of mathematics: A comparison of English and Hungarian teachers' espoused beliefs. *Journal of Curriculum Studies, 39*(3), 317–338. https://doi.org/10.1080/00220270600773082

Andrews, P., & Diego-Mantecón, J. (2015). Instrument adaptation in cross-cultural studies of students' mathematics-related beliefs: Learning from healthcare research. *Compare: A Journal of Comparative and International Education, 45*(4), 545–567. https://doi.org/10.1080/03057925.2014.884346

Aypay, A. (2010). Teacher education student's epistemological beliefs and their conceptions about teaching and learning. *Procedia - Social and Behavioral Sciences, 2*(2), 2599–2604. https://doi.org/10.1016/j.sbspro.2010.03.380

Bandura, A. (1997). *Self-efficacy: The exercise of control*. New York: Freeman.

Callejo, M., & Vila, A. (2009). Approach to mathematical problem solving and students' belief systems: Two case studies. *Educational Studies in Mathematics, 72*(1), 111–126. https://doi.org/10.1007/s10649-009-9195-z

Chan, K.-W., & Elliott, R. (2002). Exploratory study of Hong Kong teacher education students'epistemological beliefs - cultural perspectives and implications for beliefs research. *Contemporary Educational Psychology, 27*, 392–414. https://doi.org/10.1006/ceps.2001.1102

Chan, K.-W., & Elliott, R. (2004). Relational analysis of personal epistemology and conceptions about teaching and learning. *Teaching and Teacher Education, 20*(8), 817–831. https://doi.org/10.1080/0144341032000160100

Cheng, M., Chan, K.-W., Tang, S., & Cheng, A. (2009). Pre-service teacher education students' epistemological beliefs and their conceptions of teaching. *Teaching and Teacher Education, 25*(2), 319–327. https://doi.org/10.1016/j.tate.2008.09.018

Cooney, T., Shealy, B., & Arvold, B. (1998). Conceptualizing belief structures of preservice secondary mathematics teachers. *Journal for Research in Mathematics Education, 29*(3), 306–333. https://doi.org/10.2307/749792

Ernest, P. (1989). The knowledge, beliefs and attitudes of the mathematics teacher: A model. *Journal of Education for Teaching, 15*(1), 13–33. https://doi.org/10.1080/0260747890150102

Evans, J., Morgan, C., & Tsatsaroni, A. (2006). Discursive positioning and emotion in school mathematics practices. *Educational Studies in Mathematics, 63*(2), 209–226. https://doi.org/10.1007/s10649-006-9029-1

Field, A. (2005). *Discovering statistics using SPSS* (2nd ed.). London: Sage.

Gordon, S., Dembo, M., & Hocevar, D. (2007). Do teachers' own learning behaviors influence their classroom goal orientation and control ideology? *Teaching and Teacher Education, 23*(1), 36–46. https://doi.org/10.1016/j.tate.2004.08.002

Green, T. (1971). *The activities of teaching*. London: McGraw Hill.

Hannula, M. (2002). Attitude towards mathematics: Emotions, expectations and values. *Educational Studies in Mathematics, 49*(1), 25–46. https://doi.org/10.1023/A:1016048823497

Hannula, M. (2006). Motivation in mathematics: Goals reflected in emotions. *Educational Studies in Mathematics, 63*(2), 165–178. https://doi.org/10.1007/s10649-005-9019-8

Hembree, R. (1990). The nature, effects, and relief of mathematics anxiety. *Journal for Research in Mathematics Education, 21*(1), 33–46. https://doi.org/10.2307/749455

Jonsson, A.-C., Beach, D., Korp, H., & Erlandson, P. (2012). Teachers' implicit theories of intelligence: Influences from different disciplines and scientific theories. *European Journal of Teacher Education, 35*(4), 387–400. https://doi.org/10.1080/02619768.2012.662636

Lloyd, G. (2005). Beliefs about the teacher's role in the mathematics classroom: One student teacher's explorations in fiction and in practice. *Journal of Mathematics Teacher Education, 8*(6), 441–467. https://doi.org/10.1007/s10857-005-5120-2

Nosrati, M., & Andrews, P. (2016). Temporal norms of the typical mathematics lesson: Norwegian and Swedish students' perspectives. In H. Palmér & J. Skott (Eds.), *Proceedings of the 22nd mathematical views (MAVI22) conference*. Växjö, Sweden: Linnaeus University. Doi: This MAVI publication.

Op't Eynde, P., De Corte, E., & Verschaffel, L. (2002). Framing students' mathematics-related beliefs: A quest for conceptual clarity and a comprehensive categorization. In G. Leder, E. Pehkonen, & G. Törner (Eds.), *Beliefs: A hidden variable in mathematics education?* (pp. 13–37). Dordrecht, The Netherlands: Kluwer. https://doi.org/10.1007/0-306-47958-3

Op't Eynde, P., De Corte, E., & Verschaffel, L. (2006). Epistemic dimensions of students' mathematics-related belief systems. *International Journal of Educational Research, 45*(1), 57–70. https://doi.org/10.1016/j.ijer.2006.08.004

Pajares, M. F. (1992). Teachers beliefs and educational research: Cleaning up a messy construct. *Review of Educational Research, 62*(3), 307–332. https://doi.org/10.3102/00346543062003307

Rattan, A., Good, C., & Dweck, C. (2012). "It's OK - not everyone can be good at math": Instructors with an entity theory comfort (and demotivate) students. *Journal of Experimental Social Psychology, 48*(3), 731–737. https://doi.org/10.1016/j.jesp.2011.12.012

Richardson, V. (1996). The role of attitudes and beliefs in learning to teach. In J. Sikula (Ed.), *Handbook of research on teacher education* (pp. 102–119). New York: Simon and Schuster.

Schommer, M. (1990). Effects of beliefs about the nature of knowledge on comprehension. *Journal of Educational Psychology, 82*(3), 498–504. https://doi.org/10.1037/0022-0663.82.3.498

Stokking, K., Leenders, F., de Jong, J., & van Tartwijk, J. (2003). From student to teacher: Reducing practice shock and early dropout in the teaching profession. *European Journal of Teacher Education, 26*(3), 329–350. https://doi.org/10.1080/0261976032000128175

Trautwein, U., Lüdtke, O., Köller, O., & Baumert, J. (2006). Self-esteem, academic self-concept, and achievement: How the learning environment moderates the dynamics of self-concept. *Journal of Personality and Social Psychology, 90*(2), 334. https://doi.org/10.1037/0022-3514.90.2.334

Wilson, M., & Cooney, T. (2002). Mathematics teacher change and development: The role of beliefs. In G. C. Leder, E. Pehkonen, & G. Törner (Eds.), *Beliefs: A hidden variable in mathematics education?* (pp. 127–147). Dordrecht, The Netherlands: Kluwer. https://doi.org/10.1007/0-306-47958-3

The PISA Mathematics Self-Efficacy Scale: Questions of Dimensionality and a Latent Class Concerning Algebra

Boris Girnat

Abstract In 2003 and in 2012, the PISA assessment framework used a scale to measure mathematics self-efficacy. In 2015, this scale was reused in a pretest of an upcoming Swiss assessment of basic mathematical competencies in grade 9. The pretest shows three remarkable results: (1) The scale cannot be seen as unidimensional; moreover, the assumption of unidimensionality disguises some important facts, e.g. concerning gender differences. (2) The items are not worded carefully and do not seem to represent all the relevant content of mathematics adequately; concrete enhancements are suggested. (3) There are latent classes observable within the response patterns to the items, enabling to identify a latent class of "self-proclaimed algebra experts" with interesting connections to other scales measuring beliefs on mathematics.

Keywords PISA • Self-efficacy • Students' beliefs • Context questionnaire • Large-scale assessment • Mathematics education • Gender differences • Latent class analysis • Structural equation modelling • Algebra

Measuring Mathematics Self-Assessment

There are different methods for measuring pupils' self-assessment in mathematics. In general, they can be classified into two approaches: The first one is related to a person's so-called mathematical self-concept and is measured by general statements on his mathematical ability like "I have always believed that mathematics is one of my best subjects" (cf. Marsh, 1990). The second approach is called self-efficacy and is based on a suggestion of Bandura to measure a person's self-assessment not by his responses to general statements but by the level of confidence about feeling

B. Girnat (✉)
University of Applied Sciences and Arts Northwestern Switzerland School of Teacher Education, Basel, Switzerland
e-mail: boris.girnat@fhnw.ch

H. Palmér, J. Skott (eds.), *Students' and Teachers' Values, Attitudes, Feelings and Beliefs in Mathematics Classrooms*,
https://doi.org/10.1007/978-3-319-70244-5_9

able to solve specific problems that are relevant to the mathematical subdomains of interest (cf. Bandura, 1977, 1986). More explicitly, he defined self-efficacy beliefs as "people's judgments of their capabilities to organize and execute courses of action required to attain designated types of performances" (Bandura, 1986, p. 391).

Research has confirmed a correlation between mathematics self-concept, self-efficacy, and pupils' performance (cf. Multon et al., 1991), but it has been found that the first two concepts are not equivalent and that task-specific mathematics self-efficacy was even a better predictor of career choice than self-concept and test performance (Hackett & Betz, 1989). Insofar mathematics self-efficacy can be seen as a crucial part of a person's mathematical belief system (cf. Philipp, 2007, for the general concept of beliefs and Törner, 2015, for current developments). In the light of these results, scales on mathematics self-concept and self-efficacy have become essential parts of context questionnaires accompanying mathematics performance tests.

The Scales Used in PISA 2003 and 2012

The PISA studies in 2003 and in 2012 measured both the pupils' mathematics self-concept and self-efficacy using the same scales in both of these studies with one minor change (cf. OECD, 2005, pp. 291–294, & OECD, 2014, pp. 322–323). Since the mathematics self-concept is not the main focus of this article, only the eight items of the self-efficacy scale are reported here (cf. OECD, 2014, pp. 322):

Introduction: How confident do you feel about having to do the following mathematics tasks?

1. Using a train timetable to work out how long it would take to get from one place to another
2. Calculating how much cheaper a TV would be after a 30% discount
3. Calculating how many square metres of tiles you need to cover a floor
4. Understanding graphs presented in newspapers
5. Solving an equation like $3x + 5 = 17$
6. Finding the actual distance between two places on a map with a 1:10,000 scale
7. Solving an equation like $2(x + 3) = (x + 3)(x - 3)$
8. Calculating the petrol consumption rate of a car

There were four response categories, labelled with "strongly agree" (coded as 4), "agree" (3), "disagree" (2), and "strongly disagree" (1).

The items are supposed to form a unidimensional scale. Both in PISA 2003 and 2012, the Cronbach's alpha is reported. In 2003, the OECD median of Cronbach's alpha was 0.82; the Swiss value was exactly the same (OECD, 2005, p. 294). In 2012, the OECD median of Cronbach's alpha was 0.85 and 0.83 in Switzerland (OECD, 2014, p. 320). According to the usual standards of interpreting Cronbach's alpha, these values can be seen as good characteristics (cf. Cronbach, 1951).

Table 1 PISA 2003 fit statistics of a CFA model including the self-efficacy scale

Country	RMSEA	CFI	Correlation
OECD median	0.077	0.91	0.62
Switzerland	0.085	0.89	0.61

However, it is worth noting that Cronbach's alpha is only a measurement for the internal consistency of a scale, and it is not an indicator for unidimensionality. Also a scale containing (positively correlated) subscales can achieve a high internal consistency though being not unidimensional and not measuring exactly one psychological construct. An indication for the fact that exactly this situation might be instantiated by the self-efficacy scale is given in the technical report of PISA 2003. In contrast to PISA 2012, the previous documentation did not only report Cronbach's alpha but also a confirmatory factor analysis of the self-efficacy scale combined with other scales, namely, the self-concept scale and the anxiety scale. Table 1 contains the fit indices of this model and the latent correlation between self-efficacy and self-concept (cf. OECD, 2005, p. 293, & Beaujean, 2014, pp. 153–166, for interpreting the fit indices; a short summary: the RMSEA should be less than 0.06 and the CFI should be greater than 0.95, but definitively not below 0.90).

Although the PISA group states that the "model fit is satisfactory for the pooled international sample and for most country sub-samples" (OECD, 2005, p. 294), the fit indices of this model are at least on the borderline. However, since the model whose fit values are reported by the PISA group contains not only the self-efficacy scale, it is undecidable if this scale is the reason for the misfit or if one of the two other scales is responsible for the poor fit indices.

The Swiss Pretest

The Swiss Conference of Cantonal Ministers of Education initiated a nationwide assessment of basic competencies in mathematics in grade 9 (cf. EDK, 2015). This assessment is intended to take place in spring 2016. The School of Teacher Education Northwestern Switzerland is responsible for the performance test and is additionally engaged in developing a part of the context questionnaire. This questionnaire is designed in a way to be connectable with existing research. Insofar, several scales of TIMSS and PISA were integrated, and the scale of mathematics self-efficacy was of special interest. However, the results of the two PISA studies reported above give evidence for the fact that some statistical problems might occur. Since it is unclear what the reasons of these problems could be, I decided to check the PISA scale in a pretest without any changes and to revise the scale afterwards, if necessary. The pretest took place in spring 2015. It was a representative and nationwide test with 956 participants. The items of PISA 2012 were integrated into the Swiss test by using the official German, French, and Italian translations of the PISA group. In the following, I will report the results of this pretest, discussing

Table 2 Three-factor EFA of the self-efficacy scale (item 4 has been withdrawn)

Item	Factor 1	Factor 2	Factor 3	Communality	Complexity
1		0.53		0.66	1.6
2		0.68		0.63	1.1
3		0.72		0.64	1.0
5	0.83			0.90	1.3
6			0.64	0.59	1.1
7	0.94			0.87	1.1
8			0.70	0.65	1.1

what problems appeared and what I would suggest to revise this scale. After analysing the scale, I will present an interesting finding that is not based on quantitative statistics but on a latent class analysis concerning response patterns linked to the items of the scale.

Measuring Mathematics Self-Assessment

In the Swiss pretest, Cronbach's alpha was even higher than in the PISA studies having the value 0.87 with a confidence interval of [0.85, 0.89] on 95% level.

Questions of Dimensionality

As stated above, a good Cronbach's alpha does not discharge from testing the dimensionality of the scale. A parallel analysis according to Horn was performed to determine the optimal number of factors to extract (cf. Horn & Engstrom, 1979). I used the psych package (Revelle, 2015) with R (R Core Team, 2014) to process the parallel analysis and the following exploratory factor analysis (EFA). The parallel analyses suggested four factors, but the EFA showed that the fourth item "Understanding graphs presented in newspapers" caused a problem: It had a high complexity and poor and multiple loadings. That might be an evidence for the fact that the wording of this item could be misleading, e.g. it could be unclear what level of "understanding" is desired or how complex the graphs might be. After removing this item, the parallel analyses suggested three factors, and the EFA led to the following clear and simple structure (factor loadings below 0.2 are suppressed) (Table 2):

The three resulting factors can be interpreted as follows: Factor 1 is definitely the "algebra factor", whereas factor 2 and 3 can be seen as factors of applied mathematics. The difference between the latter could be located in the fact that factor 2 contains rather "easy applications", whereas factor 3 aggregates "hard applications", e.g. its items refer to tasks that require "demanding" calculations to gain the results. To confirm the explanatory outcome, the EFA was complemented by a confirmatory

Table 3 Comparison of a unidimensional and a three-dimensional CFA model

Criterion	One-factor model (14 df)	Three-factor model (11 df)
P value (χ^2)	0.000	0.474
CFI	0.971	1.000
RMSEA	0.086	0.000
SRMR	0.068	0.021

Table 4 Correlations between the three factors

	Algebra	Easy applications	Hard applications
Algebra	1		
Easy applications	0.690***	1	
Hard applications	0.552***	0.820***	1

factor analysis (CFA), using the R package lavaan (Rosseel, 2012). The three-dimensional model of the EFA was tested against the unidimensional one. In both cases, a DWLS estimator was used due to the ordinal nature of the responses (diagonally weighted least squares estimator with robust standard errors and a mean- and variance-adjusted test statistic, cf. Beaujean, 2014, pp. 92–113):

A χ^2 test indicates a significant improvement by using three factors, and the fit indices mentioned in Table 3 give also strong evidence to prefer the three-factor solution.

Correlations Between the Three Factors

In addition to the statistical values, the latent correlation between the three factors were estimated, supporting the decision in favour of the three-factor model, since especially the correlations between the algebra factor and the two application factors are too low to perceive the three factors as measuring a single psychological construct. The asterisks here and in the following denote the usual significance levels (Table 4):

Gender Differences: An Example of Practical Relevance

The discussion about dimensionality and model fits might be regarded as "purely academic", since Cronbach's alpha gives support for the operational capability of the unidimensional scale. A look on gender differences is used as an example to stress the practical relevance of these questions and to underline the warning that a questionable unidimensional scale can disguise empirical facts.

To calculate gender differences, all the latent variables are standardised, and the female group is set to be the reference group. Therefore, the female group always has zero as its mean, and the mean of the male group directly expresses the difference to the mean of the female group. Since the latent variables are standardised, the differences can be interpreted as effect sizes using the thumb rule that 0.2 indicates a small, 0.5 a medium, and 0.8 a strong effect (cf. Cohen, 1988). The gender differences are firstly calculated using the unidimensional scale (without the fourth item) and then for each of the three factors of the three-dimensional solution separately.

In case of the unidimensional scale, the group difference is 0.347** in favour of the male group. This is a small to medium effect. This finding is not unusual, but also not very remarkable (cf. Pajares, 2005). If you consider the three factors separately, the situation will change drastically: In case of the algebra factor, the group difference has a value of −0.024. This difference is not significant and practically non-existent. The difference concerning easy applications (factor 2) is small having a value of 0.276*, but the difference linked to hard applications (factor 3) rises up nearly to a strong effect of 0.766***. Insofar the unidimensional scale masks the fact that gender differences in mathematics self-efficacy is no "monolithic" issue, but it is distributed quite diversely with respect to different subdomains of mathematics.

A Proposal for Further Developments

The observation that mathematics self-efficacy can be organised in three factors leads to the question if three factors are enough to represent the mathematical content of secondary school education adequately. At least in the case of Switzerland, two domains of the traditional curriculum are not represented at all: geometry and probability. This circumstance is taken into account to reorganise the self-efficacy scale for the Swiss main test in the following manner: (1) The algebra factor and the hard applications are maintained, but each of these factors is extended to four items to broaden the possibilities of statistical investigations; (2) the easy applications are partly omitted to keep the number of items in an acceptable range; and (3) four items concerning geometry and four items concerning probability are added to represent all the relevant parts of the Swiss curriculum. The new set of items will look as follows, including as many items of the PISA scale as possible (the original PISA items are marked with an apostrophe):

1') Calculating how much cheaper a TV would be after a 30% discount
2') Calculating how many square metres of tiles you need to cover a floor
3') Calculating the petrol consumption rate of a car
4') Finding the actual distance between two places on a map with a 1:10,000 scale
5') Solving an equation like $3x + 5 = 17$
6') Solving an equation like $2(x + 3) = (x + 3)(x - 3)$
7) Developing and simplifying an algebraic expression like $2a(5a - 3b)^2$

8) Solving an equation like $2x - 3 = 4x + 5$
9) Applying the Pythagorean theorem to calculate the length of one side of a triangle
10) Constructing a perpendicular bisector using a compass and ruler.
11) Calculating the area of a parallelogram.
12) Constructing the focus of a triangle.
13) Calculating the probability of throwing a dice twice in succession to achieve two sixes.
14) Calculating the probability of getting the first prize in the lottery.
15) Calculating how likely it is to take two sweets of the same colour from a sweet jar.
16) Calculating how likely it is that two pupils in a class have the same birthday.

The purpose of these items is not only seen in representing the Swiss curriculum adequately but is also motivated by statistical reasons: The set of items contains four subsets, each of them consisting of four items. This "4x4 arrangement" is the ideal situation not only to model four independent factors but also to estimate a bi-factor model (cf. Beaujean, 2014, pp. 145–152): The entire items load on one common factor, and, additionally, each subset of four items load on one specific factor concerning applications, algebra, geometry, and probability. The common factor can be interpreted as the representation of "general" mathematics self-efficacy; the four specific factors express differences in self-efficacy related to particular subdomains of mathematics. The bi-factor model (if it will work) could fulfil two demands: primarily, the wish to measure mathematics self-efficacy in general and, secondly, the insight to take the observation seriously that it is not advisable to bundle the entire items into one unidimensional scale.

A Latent Class Analysis

A latent class analysis (LCA) is located in the qualitative or nominal part of item response theory. It uses the response patterns to items to classify the probands with a certain probability into different classes characterised by a pattern of conditional probabilities that indicate the chance that their responses to the items take on certain values (cf. Bartulucci, Bacci, & Gnaldi, 2016, pp. 81–82 and 140–141). According to the BIC criterion, the optimal number of latent classes with respect to the items of the self-efficacy scale is seven (all LCA calculations are performed by using R with the poLCA package, cf. Linzer & Lewis, 2011). Figure 1 gives a graphical overview on the probabilities of the response patterns: For each group and for each item, the red column represents the probability that a member of the respective group chooses one of the four response categories linked to this item.

The most interesting class is class 1, since its members have an extraordinarily high probability to choose the highest response option "strongly agree" with respect to the two algebra items (items 5 and 7), whereas their response probabilities to the

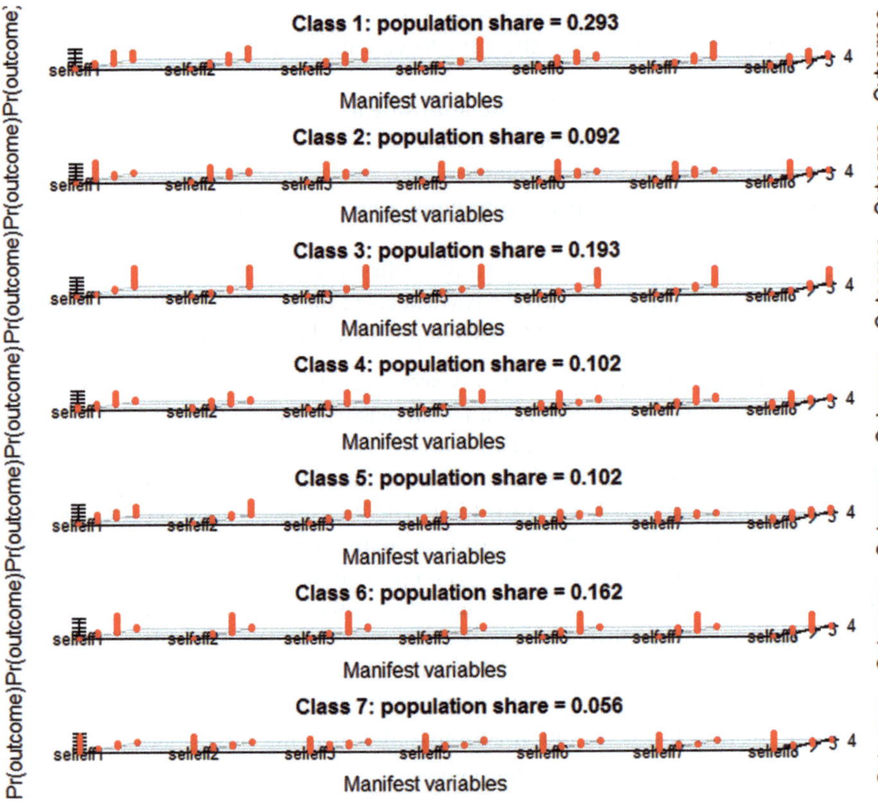

Fig. 1 Latent classes of the self-efficacy scale

other items are rather normally distributed. Insofar, this class can be labelled as the group of "self-proclaimed algebra experts".

Now, we will have a brief look on the properties of this group. Firstly, it is remarkable that 55% of the members are female, and, secondly, it is not surprising that the percentage of "algebra experts" increases according to the three school levels of Switzerland: On the lowest level, 16.1% are members of this group, 29.1% on the middle level, and 54.8% on the highest level (the latter schools are called "Gymasien", "Bezirksschulen", or "Kantonsschulen" which could be translated as academic lower secondary schools).

After considering the demographic background, I will address three topics concerning the beliefs of these pupils: The first topic is related to the self-efficacy scale, the second to preferences for teaching methods and mathematical worldviews (cf. Girnat, 2017), and the last to other scales of PISA 2012 used in the Swiss pretest like motivation, interest, and anger (cf. OECD 2014, pp. 48–66). To illustrate what these scales refer to, I will cite one item of each scale. To estimate the group differences, all the other pupils are regarded as the reference group. Insofar, the mean of the "algebra experts" can be directly interpreted as the mean difference to the other pupils.

The first topic is connected to the three subscales of the self-efficacy scale proposed above. Unsurprisingly, the group of "algebra experts" has a much higher mean on the algebra scale as the other ($d = 1.135$***), but there is just a small difference with regard to the "easy applications" ($d = 0.282$**) and remarkably a negative difference with respect to the "hard applications" ($d = -0.128$*). Insofar, the "algebra experts" do not perceive themselves as "good mathematicians" in general but only as "algebra experts", not being confident about solving "hard" mathematical applications. This finding agrees with the means of the self-concept scale ("I have always believed that mathematics is one of my best subjects"), where the difference between these two groups is not significant ($d = 0.034$).

The "algebra experts" have specific beliefs concerning preferences for teaching methods and mathematical worldviews (cf. Girnat, 2017): The system aspect ("It's necessary to understand mathematical methods. It's not enough just to apply them") is predominant ($d = 0.435$***), and also the formal aspect ("In mathematics it's important to use technical terms and conventional notations") is higher with $d = 0.330$***. Concerning the preferences for teaching methods, there is one significant contrast to other pupils: The "algebra experts" value the technique of learning by repetitive exercises higher than the others ($d = 0.464$***, "I think it's useful to solve a lot of similar tasks in order to understand a method correctly").

Finally, I will mention some group difference concerning scales adapted from PISA 2012 (cf. OECD 2014, pp. 48–66). At first, I will have a look where no significant differences could be detected. This occurs in case of the scales on anger ("I'm often that angry about my mathematics lessons that I could leave immediately"), enjoyment ("I do mathematics because I enjoy it"), instrumental motivation ("Making an effort in mathematics is worth it because it will help me in the work that I want to do later on"), and extrinsic motivation ("I want to have good marks in mathematics"). It is quite remarkable that no differences could be detected in these fields, since they might be regarded as essential to "good" performers in mathematics. The only differences that are significant could only be observed concerning two scales: intrinsic motivation ("It's important to me to understand the topics of mathematics") with $d = 0.386$*** and the learn target ("I want to learn something interesting in mathematics") with an effect size of $d = 0.362$***.

To summarise this paragraph, the "self-proclaimed algebra experts" can be seen as a relevant group of about 30% that is characterised by a special self-esteem in algebra, a specific intrinsic motivation for (the abstract and formal part of) mathematics, and a preference for repetitive techniques to learn mathematics.

Final Remarks

This article should have explained two points: Firstly, the PISA self-efficacy scale is an interesting instrument to measure pupils' mathematics self-assessment, but the scale has to be revised and cannot be regarded as unidimensional. At least according to the Swiss curriculum, a concrete enhancement of the scale was proposed, and a bi-factor model was suggested as an alternative to a unidimensional analysis.

Secondly, a latent class analysis was performed leading to the result that an interesting group of "self-proclaimed algebra experts" could be detected having specific properties related to beliefs on mathematical worldviews and the teaching and learning of mathematics: They prefer repetitive techniques to learn mathematics; and they are intrinsically interested in the formal and "theoretical" parts of mathematics, but not in its real-world applications.

The latent class analysis stresses a possibly unintended advantage of Bandura's concept of self-efficacy: The competitive approach of a person's mathematical self-concept based on general statements on mathematics self-assessment would not be suitable to detect different response patterns related to diverse subdomains of mathematics. Insofar, Bandura's concept offers possibilities being interesting both with regard to contents and statistical methods: The latent class analysis and the bi-factor model suggested above allow a more fine-grained investigation of pupils' performance-related beliefs than the mathematical self-concept. But this statement is not to be interpreted as an advice to replace the mathematics self-concept by self-efficacy in general. As shown above, both approaches can complement one another: The "self-proclaimed algebra experts" only have got a higher self-esteem concerning algebra (measured by self-efficacy scales) and not concerning mathematics in general (measured by self-concept scales). Insofar, a combination of both approaches is an opportunity to detect subtle properties of pupils' beliefs related to their own mathematical performance and potential.

References

Bandura, A. (1977). Self-efficacy: Toward a unifying theory of behavioral change. *Psychological Review, 84*, 191–215. https://doi.org/10.1037//0033-295x.84.2.191

Bandura, A. (1986). *Social foundations of thought and action: A social cognitive theory*. Englewood Cliffs, N.J: Prentice Hall. https://doi.org/10.4135/9781446221129.n6

Bartulucci, F., Bacci, S., & Gnaldi, M. (2016). *Statistical analysis of questionnaires – A unified approach based on R and Stata*. Boca Raton, FL: CRC Press.

Beaujean, A. (2014). *Latent variable modeling using R – A step-by-step guide*. New York: Routledge.

Cohen, J. (1988). *Statistical power analysis for the behavioral sciences* (2nd ed.). Hillsdale, NJ: Lawrence Erlbaum Associates.

Cronbach, L. J. (1951). Coefficient alpha and the internal structure of tests. *Psychometrika, 16*, 297–334. https://doi.org/10.1007/bf02310555

EDK (Swiss Conference of Cantonal Ministers of Education). (2015). *Faktenblatt Nationale Bildungsziele (Grundkompetenzen): Definition, Funktion, Überprüfung* (fact sheet national goals of education (basic competencies): definition, purpose, assessment) http://www.edudoc.ch/static/web/arbeiten/harmos/grundkomp_faktenblatt_d.pdf

Girnat, B. (2017). Gender differences concerning pupils' beliefs on teaching methods and mathematical worldviews at lower secondary schools. In C. Andrà, D. Brunetto, E. Levenson, & P. Liljedahl (Eds.), *Teaching and learning in maths classrooms: Emerging themes in affect-related research: teachers' beliefs, students' engagement and social interaction (Research in Mathematics Education)* (pp. 253–263). Cham: Springer International Publishing AG, S. https://doi.org/10.1007/978-3-319-49232-2_24)

Hacket, G., & Betz, N. (1989). An exploration of the mathematics efficacy/mathematics performance correspondence. *Journal of Research in Mathematics Education, 20*, 261–273. https://doi.org/10.2307/749515. National Council of Teachers of Mathematics, Reston.

Horn, J. L., & Engstrom, R. (1979). Cattell's scree test in relation to bartlett's chi-square test and other observations on the number of factors problem. *Multivariate Behavioral Research, 14*(3), 283–300. https://doi.org/10.1207/s15327906mbr1403_1

Linzer, D. A., & Lewis, J. B. (2011). poLCA: An R package for polytomous variable latent class analysis. *Journal of Statistical Software, 42*(10), 1–29. URL http://www.jstatsoft.org/v42/i10/. https://doi.org/10.18637/jss.v042.i10

Marsh, H. W. (1990). *Self-description questionnaire (SDQ) II: A theoretical and empirical basis for the measurement of multiple dimensions of adolescent self-concept: An interim test manual and a research monograph.* San Antonio, TX: The Psychological Corporation.

Multon, K. D., Brown, S. D., & Lent, R. W. (1991). Relation of self-efficacy beliefs to academic outcomes: A meta-analytic investigation. *Journal of Counselling Psychology, 38*., American Psychological Association, Washington, D.C, 30–38. https://doi.org/10.1037/0022-0167.38.1.30

OECD. (2005). *PISA 2003 technical report*. Paris, France: OECD.

OECD. (2014). *PISA 2012 technical report*. Paris, France: OECD.

Pajares, F. (2005). Gender differences in mathematics self-efficacy beliefs. In A. M. Gallagher & J. C. Kaufman (Eds.), *Gender differences in mathematics – An integrative psychological approach* (pp. 294–315). Cambridge, UK/New York: Cambridge University Press. https://doi.org/10.1017/cbo9780511614446.015

Philipp, R. A. (2007). Mathematics teachers' beliefs and affect. In F. K. Lester (Ed.), *Second handbook of research on mathematics teaching and learning* (pp. 257–315). Charlotte, NC: Information Age Publishing.

R Core Team. (2014). *R: A language and environment for statistical computing*. R Foundation for Statistical Computing, Vienna, Austria. URL http://www.R-project.org/

Revelle, W. (2015). *Psych: Procedures for personality and psychological research*. Evanston, IL: Northwestern University. http://CRAN.R-project.org/package=psych Version = 1.5.8.

Rosseel, Y. (2012). Lavaan: An R package for structural equation modeling. *Journal of Statistical Software, 48*(2), 1–36. URL http://www.jstatsoft.org/v48/i02/, Version = 0.5-20 from https://cran.r-project.org/package=lavaan. https://doi.org/10.18637/jss.v048.i02

Törner, G. (2015). Beliefs – No longer a hidden variable, but in spite of this: Where are we now and where are we going? In L. Sumpter (Ed.), *Current state of research on mathematical beliefs XX: Proceedings of the MAVI-20 conference September 29–October 1, 2014* (pp. 7–20). Falun, Sweden: Högskolan Dalarna.

The Influence of Assessment on Students' Experiences of Mathematics

Helena Roos

Abstract The empirical material and results presented in this paper come from an ongoing ethnography-inspired study of inclusion in mathematics as seen from a student perspective. This study did not initially focus on assessment, but when investigating what influences students' experiences of school mathematics, assessment came out as a result. The research participants are not ordinary students, but students who need some degree of special education in mathematics, either as gifted or as low-performing students. For these students, traditional assessment in mathematics does not provide any relevant feedback to support them. On the whole, assessment primarily influences either how they write solutions to tasks, but not exactly how they solve them, or else how they feel about themselves as low performers in mathematics.

Keywords Special educational needs in mathematics • Assessment • Discourse analysis

Introduction

An important question in mathematics education asks how students experience mathematics in school and what influences how they develop mathematics knowledge. In response, the main purpose of this paper is to investigate students' experiences and perceptions of mathematics at school. The students investigated are not any ordinary students, but ones who need some degree of special education in mathematics (SEM), either as gifted or as low-performing students.

H. Roos (✉)
Linnaeus University, Växjö, Sweden
e-mail: Helena.roos@lnu.se

H. Palmér, J. Skott (eds.), *Students' and Teachers' Values, Attitudes, Feelings and Beliefs in Mathematics Classrooms*,
https://doi.org/10.1007/978-3-319-70244-5_10

SEM is about students in need of something else other than what is typically offered in mathematics education in order to be able to optimise their learning of mathematics (Magne, 2006). In this study, SEM is identified as an educational initiative that may occur if a student is a high or a low achiever either in general or in specific areas of mathematics. In this paper, the phrase *student in need of SEM* is used to imply that a student is *in* need of something from his or her surroundings in order to appropriately develop his or her mathematical knowledge (Bagger & Roos, 2015).

Across and even within countries, schools deal differently with SEM. One way is to work in inclusive settings (Persson & Persson, 2012), although that method invariably raises the question of what working inclusively means exactly. From a broad perspective, working inclusively means being able to accommodate all differences among students within normal classrooms and create opportunities for all students to participate meaningfully in their education (Barton, 1997; Persson & Persson, 2012). Its aims are to organise schools around the fact that students are different and to ensure every student's participation in relevant learning activities (Nilholm, 2006). Accordingly, participation is an important aspect of inclusive education, yet one that invariably raises another question: Participation in what?

This study of inclusion in mathematics as seen from the perspectives of the students was conducted in a lower secondary school in Sweden. Its aim is to investigate what these students perceive themselves to be participating in and how that perception influences their experiences with mathematics in school. Their experiences are described as discourses, among which the discourse of assessment is the primary focus of the paper.

Assessment

This paper concerns how assessment influences students' experiences of mathematics in school. The Swedish primary school curriculum (Swedish National Agency, 2011) stipulates that teaching needs to be examined and evaluated. Another aspect of evaluation in Swedish schools is assessment of students' knowledge and the responsibility to inform students, parents, and school principals about the knowledge development of individual students. The Swedish National Agency for Education provides both oral and written tests, the stated purpose of which is to provide a standard assessment of pupils and increase the achievement of learning goals.

Of course, assessment is a broad, complex concept, and many situations in a student's school day can be regarded as forms of assessment, such as day-to-day communication with teachers, tests, and class work (Björklund-Boistrup, 2010). In any case, assessment can come in two types: formative and summative. On the one hand, *formative assessment* encompasses all 'activities undertaken by teachers, and/or by their students, which provide information to be used as feedback to modify the teaching and learning activities in which they are engaged' (Black & Wiliam, 1998, p. 7). On the other, *summative assessment* is performed with tests on a local or national level and summarises students' performance in relation to stated goals (Björklund-Boistrup, 2010).

Lyon (2011) explains that research on assessment at the classroom level usually takes an assessment-, teacher-, or student-centred approach. The most studied of those three types is the assessment-centred approach, whereas the student-centred approach, which focuses on motivation and achievement, receives less attention (Lyon, 2011; Wiliam, 2007). Boaler (1998) investigated mathematics assessment from a student-centred approach to show that the form of assessment influenced student's knowledge and the ways in which students applied school-learned mathematical methods in situations outside the mathematics classroom. In this paper, the student-centred approach is applied to examine how assessment influences students' experiences of mathematics in school.

Theoretical and Methodological Perspectives

This study has used discourse analysis (DA). Although some research applies DA as an analytical tool only, other research applies it as a theory, while still other studies, such as this one, apply it as both. Common to all of those approaches is a focus on language and text: what we can actually see, hear, and read. Such a focus can be applied with different approaches, since the field of DA makes many routes of application available. Similar to all approaches, however, is that DA concerns studying language in use and examining language beyond its use in sentences (Trappes-Lomax, 2006)—that is, the meaning of language in interaction. By extension, from a DA perspective, when we create texts, we are active reproducers of culture (Gee, 2005). One way to capture students' perspectives on inclusion in mathematics education is therefore to grasp how students perceive the ways in which they are included in mathematics taught in different situations, which can be done by identifying the ways in which students talk about, act, and produce items in school mathematics.

This study applies DA as described by Gee (2005, 2011). From his perspective, DA encompasses all forms of interaction, both spoken and written, and he provides a toolkit for analysing such interaction. The toolkit consists of 28 tools of inquiry and stresses the fact that speakers and writers are active designers in reproducing culture (Gee, 2011). Although limited space prevents the tools from being fully described here, they generally focus on communication and ask questions of texts.

Gee (2005, 2011) also distinguishes two theoretical notions, big and small discourses, henceforth referred to as *Discourse* and *discourse*, respectively. On the one hand, Discourse represents a wider context, both social and political, and is constructed upon ways of saying, doing, and being. In any case, recognition is critical. Such Discourses are always simultaneously embedded in various social institutions involving various sorts of properties and objects. For example, Discourse can be mathematics in school, although any Discourse encompasses language plus 'other stuff' (Gee, 2005, p. 52), including actions, interactions, values, symbols, objects, tools, and places. When language and the other stuff are combined in a way that makes them recognisable, the result is Discourse, and the persons engaged in that Discourse are recognised as a particular type. On the other hand, discourse focuses on language in use: the 'stretches of language' that we can see in conversations or stories that we investigate (Gee, 2011). In this study, Discourses and discourses inform the theoretical perspective, which is applied by using the abovementioned toolkit as a methodological instrument.

Text was analysed with the help of the tools of inquiry provided by Gee (2011). While examining the text, I asked certain questions depending on the type of text being examined. For example, when using the subject tool, I asked, 'What are they talking about here, and why?' When using the deictic tool, I asked, 'What is pointed out in the text, and what is the listener assumed to already know?' When applying the fill-in tool, I asked, 'What needs to be filled in to achieve clarity? What is not being said overtly, but is nevertheless assumed to be known or inferable?'

When students were addressing the same aspects, different themes emerged—or to use Gee's (2011) terminology, different 'stretches' of language appeared. Thereafter, Discourses were identified by how the speakers constructed the stretches of language and if such stretches could be seen in classroom observations and, if so, then how.

This study also draws upon ethnography, meaning that the researcher has sought to understand a phenomenon through interpersonal methods over time by collecting data via social interactions (e.g., interviews, discussions, and visual representations). Social interactions (Aspers, 2007) and in-depth studies (Hammersley & Atkinson, 2007) prescribed by ethnography can be used to follow certain processes in a particular case. In this study, ethnography was applied together with DA in order to make students' perspectives of mathematics teaching and learning visible. However, it is important to recognise that conflicts can arise when using ethnography and DA together (Hammersley, 2005). Such conflicts can emerge both in ways of conducting empirical investigations and in the roles, if any, that researchers play in their research. Nevertheless, in this research, DA and the ethnographic approach complement each other; DA provides theoretical and analytical notions, while ethnography provides a methodology for conducting research.

Setting the Scene

I examined a municipal lower secondary school (Grades 7–9) in Sweden that has roughly 500 students and that has set out to implement inclusive work. The goal of applying inclusive work implies that the school does not typically apply special education in small groups, that all teaching occurs in the regular classroom, and that there are nearly always two teachers for each lesson—for mathematics lessons, one SEM teacher and one regular mathematics teacher. The school is an urban one, albeit on the outskirts of a city, and has a varied catchment area of both apartment blocks and villa districts. As such, the students examined have different socioeconomic backgrounds.

Two classes (i.e. Grade 7 and Grade 8) selected by the school's teachers were examined via observations and interviews. The classes were selected according to how long the students had been participating in inclusive classrooms. In Grade 7, inclusive settings were rather new, whereas in Grade 8, students had engaged in inclusive learning for at least a year and planned to continue it for another year. Another criterion was that the classes could handle having a researcher in the classroom. Teachers also identified students in need of SEM in the two classes based on the definition presented in the introduction; their selection included both students struggling with mathematics and students who needed additional learning challenges.

Since this research focuses on students and special education needs, ethical considerations were taken into account before, during, and after the research process. Both the students and their guardians have provided their written consent. Furthermore, as the researcher, I reflected on the ways in which I could have affected the students and the study. Above all, I did not want to put students in any uncomfortable situations or make them feel exposed. Floyd and Arthur (2012) highlight the importance of researchers' being aware of both external and internal ethics. Here, the external ethical issues were the visible aspects—for example, written consent—whereas the internal ones related to my possible ethical and moral dilemmas as a researcher in relation to the research conducted. In that sense, preventing students' exposure and my being an unfamiliar adult in their classrooms were important to consider.

Several methods were used for data collection. All students involved completed a written self-evaluation. Observations of mathematics lessons were conducted in order to study the context, and after the observations close in time student interviews were conducted. For this paper, two students—Edward and Ronaldo—were selected as examples because they both attended the same class, but Edward is perceived to be a gifted student in mathematics, whereas Ronaldo is perceived to be a low performer. The students chose their pseudonyms.

Edward and Ronaldo

Edward and Ronaldo are both 15 years old and in the same class in year 8. They also both attend the same mathematics class, have the same mathematics teachers, and, by that, receive 'the same' mathematics teaching. Edward is perceived to be a gifted student in mathematics and has earned A marks, the highest mark on a scale from A (high aptitude) to F (fail) in which A–E marks are passing. By contrast, Ronaldo is perceived as a low performer in mathematics and has earned roughly E marks. The special education teacher in mathematics describes the two boys as follows:

> Edward needs more challenges. He cracks the codes and such things at an entirely different level from the rest of the class. So, he needs to be challenged. He has a pretty clear focus of what he wants in the future. He wants to be prepared for upper secondary school, so we [the teachers] must be better at challenging him.

> He [Ronaldo] has an approved grade, but we feel that it's probably pretty much thanks to the adjustments that he receives . . . adapted materials, he gets introductions of lessons and stuff in a small group . . . . He gets a lot of directed support, which has enabled him to reach the [educational] goals.

For this paper, the first two interviews with each of Edward and Ronaldo were analysed using of DA (Gee, 2011). The four interviews were conducted by the same researcher at the school in a small room next to the classroom. The first interview was based on a written self-assessment that the students completed and that contained claims about how they perceive their mathematics teaching and learning—for example, how they feel (i.e. *sure*, *pretty sure*, *unsure*, or *very unsure*) when they are going to tell a friend how they have solved a task or when they are going to choose a method to solve a task. The second interview was based on a task used in a previous lesson about the circumference of a bicycle wheel. The focus of the results is assessment, even though assessment was not expressly addressed in the self-assessments or interview questions. On the contrary, the students introduced assessment as a topic in the interviews.

Analysis and Results

In the four interviews, stretches of language-addressing assessment appeared several times, although assessment was not an expressed focus of the interview questions. These stretches are presented in what follows.

In the first interview, Edward was asked about how he writes in mathematics:

Edward: Well, it mostly takes place in my head, but then when it's a test, you write everything.

Interviewer: Okay. Why do you do it [write] then [on the tests]?

Edward: Otherwise. .. you cannot be assessed on what you have done, but when I do the calculations in my workbook, then it is mostly mental arithmetic.

Later in the interview, Edward described an ordinary mathematics lesson.

Edward: They [the lessons] are kind of good. [...] well, they [the teachers] have lesson introductions on the blackboard, then we [the students] are supposed to do the calculations in the book. And then on the blackboard, they do different E, C, and A tasks. Or not [A tasks], but they make up C [tasks] on the board anyway.

Interviewer: And then you mean the grading [A, C, and E]?

Edward: Yes, the grade levels, because they show approximately how difficult the task we did [on the blackboard] was.

Interviewer: mm ... ah, right, would you like it to be done any differently?

Edward: No, it's too complicated to pick up an A task on the blackboard, because it's so much to write, and often, it's problem solving.

Here, Edward referred to how he performs arithmetic mostly in his head. However, when he takes a test, he writes out his calculations on paper so that he can be assessed, even though when he has a regular math lesson, he mostly uses mental arithmetic. If using the fill-in tool, Edward thinks that he has to do mathematics differently in different situations, and he introduces assessment into the discussion, thereby making assessment a topic that was not expressly addressed. He also described the mathematics lesson from the perspective of assessment, in which he refers to grades to describe the different tasks done on the blackboard by teachers. When referring to an A task, he does not say that he wants an A task on the blackboard, but he indirectly says that he wants a more difficult task. Although A tasks are not defined, they are often referred to as problem-solving and tasks that need to be explained with 'so much' writing.

In Ronaldo's first interview, he discussed taking tests:

Ronaldo: Well, it always feels pretty good when I take the tests, but then … it becomes a little like—when you get the result and then I think that it will get better next time. Like struggling, like struggle more and more like that ….

Interviewer: mm .. How was it then, this test?

Ronaldo: Well, I got like one point away from a D or something ...

Interviewer: But it did not feel good?

Ronaldo: Well, yeah ….

When Ronaldo talked about 'it' here, he referred to how he generally feels about assessment and how well he performs on tests. He also used vocabulary that relates to his feelings; he says that he will be 'struggling more', and when he talks about how it feels, his intonation is hesitant and uses ambiguous language—'Well, yeah'—which indicates that he does not feel exactly pleased about the result of the test, even though the question was more or less guiding him towards a positive answer.

In the second interview, the focus was a geometry lesson in which the students needed to calculate the circumference of a bicycle wheel and then how far the bicycle had gone when the wheel had made 1000 revolutions. I observed the lesson and conducted the interviews a few days later. As shown in the following, Edward explained his thinking when facing the task and explained that although he does not need to perform all of the small steps in the calculations, he feels as if he has to show how he performs them anyway.

Edward: It's just that you have to do it [write out the steps].

Interviewer: To show?

Edward: Really, it's just a burden to do that.

Interviewer: (*Laughing.*)

Edward: (*Laughing.*) To make and write all that, because it takes such a [long] time.

Interviewer: Otherwise?

Edward: Otherwise I just do it so quickly.

Interviewer: So you really don't have to do all of those calculations?

Edward: No, no. I would have done it in very few calculations.

[. . .]

Interviewer: But you can't not do that?

Edward: No, I can't.

Interviewer: Because?

Edward: Then I can't be assessed.

[. . .]

Edward: It's that you have to do it on the tests or you won't pass. But in the math book, I don't do it. . . . It's the writing that takes such a long time on tests.
Interviewer: Yes, that's right.
Edward: Because I take the entire time [when taking tests].
Interviewer: You do?
Edward: To *have time* to write everything.
Interviewer: Ah.
Edward: I mean, pure physically.

In this excerpt, Edward discussed having to write out calculations. He used the words 'have to' and 'burden', thereby indicating that he does not need or want to write out the mathematical steps. He also talked about mental calculations versus written ones and said that writing them out 'takes such a long time', but mentally, he can do the arithmetic 'just like that'. However, he has to write out the steps or else he 'cannot be assessed'. Edward especially stressed the burden of writing everything on tests by emphasising that he is given time to do so. He also pointed out that it is not a burden mentally, only 'physically'.

Also in the second interview, Ronaldo discussed completing geometry tasks in the mathematics book:

Ronaldo: I think that Level 1 [referring to tasks at Level 1 in the book] has begun to be a little easier now, because it feels like I've started to get a little better at maths. So I start, like, with Level 2.
Interviewer: You have started at the second level now?
Ronaldo: Yes, and then Level 3.
Interviewer: Nice. How does it feel?
Ronaldo: Good. Like progress.

Here, Ronaldo addressed the levels of tasks in the mathematics book, among which Level 1 has 'easy tasks while the tasks at Level 4 provide real challenges', according to the book. He referred to his feelings about the levels of the tasks and indirectly said that to be good at maths, one has to be able to do the higher-level tasks. He assessed himself in relation to the levels of tasks, and since starting with Level 2 tasks, he has become a 'little better at maths'. He used the word 'progress' to describe the feeling.

Taken together, the discourses in the analysed interviews exhibit strong stretches of assessment in mathematics introduced directly by the students and not the interview questions or interviewer. Arguably, the stretches about assessment indicate the Discourse of assessment in mathematics, for the students indeed position themselves as being assessed in mathematics, which clearly influences their experiences of the subject. Interestingly, the students are categorised by their teachers as either gifted or low performing in mathematics, yet both are greatly influenced, albeit differently, by the Discourse of assessment in mathematics. On the one hand, assessment influences Edward at the level of actions, such as his writing out solutions to tasks in a way that will earn him good grades. That way of writing is not how he actually solves the tasks but merely a construction for assessment; as such, tension exists between writing solutions for him and for others. On the other hand, Ronaldo is influenced by assessments at the level of feelings, particularly his feelings about himself as someone who is not very good at mathematics. However, it is premature to say whether the differences relate to the different needs of the two students—one gifted and one low performer—or not.

Discussion

Mathematics teaching in Swedish primary schools varies considerably, which creates differences in students' experiences of the subject and how it is taught (Swedish Schools Inspectorate, 2009). The different social and cultural contexts within which children learn mathematics influence what they learn, what they think mathematics is, and how they think about mathematics learning (Perry & Dockett, 2008). Based on the results presented in this paper, assessment seems central to both Edward's and Ronaldo's experiences of mathematics at school, but in different ways, as seen in how they talk and reproduce culture (Gee, 2005). Consequently, assessment is a Discourse instead of a discourse. On the one hand, Edward is influenced by the Discourse by feeling as though he has to write out even the smallest of steps in solving mathematical problems in order to be assessed well. For him, doing so is a burden that he nevertheless assumes because he has learned that not doing it is to his disadvantage in assessment. This result indicates that assessment and learning do not always go hand in hand. Edward also indirectly indicated that he wants to discuss more difficult tasks during mathematics lessons but that the form of A tasks is too challenging, thereby suggesting that he needs another level of mathematics to be challenged in assessments. Ronaldo, on the other hand, is influenced by assessment at a level of feelings. When he got one point away from a D mark, he said that he feels as if he struggles more, thereby indicating he is not pleased with his result. He also talked about his knowledge in mathematics in relation to levels in the mathematics book, thereby suggesting he is assessing his own knowledge in relation to the levels and that he has made progress since beginning at a higher level.

In general, it is apparent that assessment affects students' experiences of mathematics. In this case, written assessments (i.e. tests) were the most obvious means of assessment to students, even though many other situations involved the assessment of students' knowledge in mathematics (Björklund-Boistrup, 2010). Formative assessment was not visible in the interviews, meaning that the students did not view other situations as being assessments, but only the written tests.

In sum, it seems as though assessment does not provide relevant input to support the SEM students in this study in developing their mathematical knowledge, as the Swedish National Agency (2011) prescribes, to increase students' achievement of goals. On the contrary, assessments seem to exert other kinds of influence on the students: in Ronaldo's case, his feelings about mathematics and how he perceives himself to be a low performer, whereas in Edward's case, the burden he feels about having to present his knowledge. In that sense, it seems that assessment is even an obstacle for Edward, a gifted student, in developing his knowledge of mathematics.

References

Aspers, P. (2007). *Etnografiska metoder: att förstå och förklara samtiden. [Ethographical methods: to understand and explain contemporary life]* Malmö: Liber. ISBN: 978-91-47-08648-1.

Bagger, A., & Roos, H. (2015). How research conceptualises the student in need of special education in mathematics. I O. Helenius, A. Engström, T. Meaney, P.Nilsson, E. Norén, J. Sayers, & M. Österholm (red.), *Development of mathematics teaching: Design, scale, effects.* (pp. 27–36). Proceedings from MADIF9: The Ninth Swedish Mathematics Education Research Seminar, Umeå, February 4–5, 2014. Linköping: SMDF. ISBN: 978-91-973934-9-2.

Barton, L. (1997). Inclusive education: Romantic, subversive or realistic? *International Journal of Inclusive Education, 1*(3), 231–242. https://doi.org/10.1080/1360311970010301

Björklund-Boistrup, L. (2010). *Assessment discourses in mathematics classrooms: A multimodal social semiotic study [Doctoral dissertation].* Stockholm: Stockholm University. ISBN: 978-91-7447-116-8.

Black, P., & Wiliam, D. (1998). Assessment and classroom learning. *Assessment in Education: Principles, Policy & Practice, 5*(1), 7–73. Englewood Cliffs, NJ: Prentice Hall. https://doi.org/10.1080/0969595980050102

Boaler, J. (1998). Alternative approaches to teaching, learning and assessing mathematics. *Evaluation and Programming. Planning, 21*, 129–141. https://doi.org/10.1016/S149-7189(98)00002-0

Floyd, A., & Arthur, L. (2012). Researching from within: External and internal ethical engagement. *International Journal of Research & Method in Education, 35*(2), 1–10. https://doi.org/10.1080/1743727X.2012.670481

Gee, J. P. (2005). *An introduction to discourse analysis: Theory and method.* New York, NY: Routledge. ISBN: 978-415-72125-7.

Gee, J. P. (2011). *How to do discourse analysis: A toolkit.* London, UK: Routledge. ISBN: 978-0-415-72557-6.

Hammersley, M. (2005). Ethnography and discourse analysis: Incompatible or complementary? *Polifonía, 10*(10), 1–19.

Hammersley, M., & Atkinson, P. (2007). *Ethnography: Principles in practice* (3rd ed.). London, UK: Routledge. 978-0415396059.

Lyon, E. G. (2011). Beliefs, practices, and reflection: Exploring a science teacher's classroom assessment through the assessment triangle model. *Journal of Science Teacher Education, 22*, 417–435. https://doi.org/10.1007/s10972-011-9241-4

Magne, O. (2006). Historical aspects on special education in mathematics. *Nordic Studies in Mathematics Education, 11*(4), 7–34.

Nilholm, C. (2006). *Inkludering av elever 'i behov av särskilt stöd': Vad betyder det och vad vet vi? [Incluision of students in need of 'special support'; what does it mean and what do we know?] Forskning i Fokus 28.* Stockholm, Sweden: Myndigheten för skolutveckling. ISSN:1651-3460.

Persson, B., & Persson, E. (2012). *Inkludering och måluppfyllelse: Att nå framgång med alla elever [inclusion and goal achievement: To reach success with all students].* Stockholm, Sweden: Liber. ISBN: 9789147100880.

Perry, B., & Dockett, S. (2008). Young children's access to powerful mathematical ideas. In L. D. English (Ed.), *Handbook of international research in mathematics education* (pp. 75–108). New York, NY: Routledge. ISBN: 978-0415832045.

Swedish National Agency. (2011). *Läroplan för grundskolan, förskoleklassen och fritidshemmet 2011. [Curriculum for elementary school, pre-school-Class, and leisure Centre 2011].* Stockholm, Sweden: Fritzes. ISBN: 978-91-38325-42-1.

Swedish Schools Inspectorate. (2009). *Undervisningen i matematik: Utbildningens innehåll och ändamålsenlighet [teaching mathematics: The content and effectiveness of education]. Kvalitetsgranskning rapport 5.* Stockholm, Sweden: Skolinspektionen.

Trappes-Lomax, H. (2006). Discourse analysis. In A. Davies & C. Elder (Eds.), *The handbook of applied linguistics* (pp. 133–164). Malden, MA: Blackwell. ISBN: 9780631228998.

Wiliam, D. (2007). Keeping learning on track: Classroom assessment and the regulation of learning. In F. K. Lester (Ed.), *Second handbook of research on mathematics teaching and learning* (pp. 1051–1098). Charlotte, NC: National Council of Teachers of Mathematics & Information Age. ISBN: 978-1593115869.

Gjennomgang and Genomgång: Same or Different?

Paul Andrews and Mona Nosrati

Abstract In this paper, we examine whether two words, gjennomgang and genomgång, loosely translated as "going through", represent common didactical practices in Norwegian and Swedish upper secondary mathematics classrooms. Data from semi-structured group interviews yielded students' perceptions as a belief synthesis of many years' experience of mathematics classrooms. Analyses indicated that Norwegian vocational students experience a directive "going through" during which teachers inform them what work they will be doing from the book or computer. The remaining students described two forms of "going through": instructive "going throughs" whereby teachers model new procedures and problem-solving "going throughs", in which teachers demonstrate solutions to problems that students had previously found difficult.

Keywords Student beliefs • Genomgång • Gjennomgang • Mathematics • Norway • Sweden • Upper secondary school • Whole class teaching

Introduction

This serendipitously motivated paper draws on data from semi-structured interviews during which Norwegian and Swedish upper secondary students were asked to describe a typical mathematics lesson, where upper secondary school refers to post-compulsory education offered to students typically 16 years and older. During our analyses, we noticed students referring to a form of classroom activity as a gjennomgang (Norwegian) or genomgång (Swedish) in ways that implied a tacit acceptance by both teachers and students of the words' meaning. However, both words were unfamiliar to Andrews, a relatively recent immigrant to Sweden, prompting the question, are gjennomgang and genomgång, translated as "going through",

P. Andrews (✉)
Stockholm University, Stockholm, Sweden
e-mail: paul.andrews@mnd.su.se

M. Nosrati
NTNU, Trondheim, Norway

H. Palmér, J. Skott (eds.), *Students' and Teachers' Values, Attitudes, Feelings and Beliefs in Mathematics Classrooms*,
https://doi.org/10.1007/978-3-319-70244-5_11

construed similarly as commonly enacted didactical traditions? Embedded in such a question lies an assumption, supported by a growing literature, that the ways in which teachers teach and students learn are culturally normative. For example, while teachers in England, Flanders, Hungary and Spain have been observed to spend similar proportions of lesson time on the public explanation of mathematical ideas, the extent to which they posed high-level questions for public response or invited students to share publicly their problem solutions varied considerably (Andrews, 2009a). In similar vein, a second analysis of the same data showed that while teachers encouraged the development of both conceptual knowledge and procedural knowledge, there was considerable cross-cultural variation with respect to learning outcomes like problem solving or the development of students' mathematical reasoning (Andrews, 2009b).

Furthermore, while research shows that the contexts in which they operate, the role of their teachers, parents and friends and day-to-day classroom interactions influence both the formation and manifestation of students' mathematics-related beliefs, few studies have examined what students believe happens during their lessons. Admittedly, some studies have indirectly examined the phenomenon (Op't Eynde, De Corte, & Verschaffel, 2006), but they used literature-derived surveys rather than seek directly students' views as to the nature of the teaching they receive, an omission this paper aims to redress. Moreover, with respect to warranting this paper, we note that Häggström (2006), a cultural insider, did not mention genomgång when discussing the ways in which Swedish teachers introduce their mathematics lessons, an omission yielding at least three possible consequences. Firstly, genomgång is not the common occurrence our interviewees implied. However, conversations with teacher education colleagues across the Nordic world indicate that while no such word exists in Finland, genomgång (Sweden), gjennomgang (Norway) and gennemgang (Denmark) are commonplace. Secondly, cultural outsiders frequently notice routines hidden from the cultural insider, indicating that genomgång and gjennomgang may be "folkways" of Scandinavian teaching (Buchmann, 1987). That is, while such practices may be educationally effective, they are typically uncodified and "warranted by their existence and taken-for-granted effectiveness" (Buchmann, 1987, p. 154). In this sense, folkways "are learned by tradition and imitation; having the authority of custom and habit... insiders feel that the folkways are 'true' and 'right'" (Buchmann, 1987, p.155). Indeed, the same conversations with teacher education colleagues indicated that these nationally located forms of "going through" are not formally addressed in teacher education courses. Thirdly, when attending school, students unconsciously assimilate the mathematics teaching practices common to their country's lessons, typically because the routine enactment of culturally normative didactical practices (Andrews & Sayers, 2013) shapes how students perceive them (Knipping, 2003). In particular, when participants' speak of gjennomgang and genomgång, they seemed to us to be drawing on collective "mental images" of what teaching typically looks like (Stigler & Hiebert, 1999). Such matters frame our interests in what Norwegian and Swedish upper secondary students believe about gjennomgang and genomgång, respectively.

However, in presenting the study as an examination of what students believe, it is necessary to discuss how research has construed beliefs. Indeed, a secondary goal of this paper, which we address it in the discussion, is to examine the extent to which students' interview utterances can be construed as beliefs.

On Beliefs: Their Form

In the following, we remind ourselves of some of the oft-neglected earlier literature on beliefs and examine their significant and generally agreed forms and functions. With respect to their form, it is widely agreed that beliefs offer representations of reality that inform subsequent actions (Brown & Cooney, 1982; Ernest, 1989), although the extent to which causality can be inferred may be contextually limited (Skott, 2009). In broad terms, beliefs form clusters or systems focused on particular phenomena in ways that allow for an individual to hold apparently conflicting beliefs (Abelson, 1979; Nespor, 1987). The beliefs within a system, which can be either primary or derivative, are typically located somewhere between the system's centre and periphery, with primary beliefs located at the centre of the system being least susceptible to influence (Green, 1971).

Importantly, particularly with respect to distinguishing them from knowledge, beliefs are both non-consensual and unbounded (Abelson, 1979; Nespor, 1987). They are non-consensual in the sense that "there are no clear logical rules for determining the relevance of beliefs to real-world events and situations" (Nespor, 1987, p.321). By way of example, Abelson invites the reader to:

> Consider some societal problem area like, say, the Generation Gap. Youngsters may have a highly articulated system of concepts blaming the problem on adult restrictiveness and insensitivity, whereas oldsters develop concepts around adolescent rebellion and immaturity. Meanwhile, psychologists may view the matter in terms of communication failure between generations. (Abelson, 1979, p.356)

Indeed, the variation in the three group's hypothetical perspectives shows well how the belief construction requires no consensus. This idiosyncratic and highly personal nature of belief construction alludes to their unboundedness. Here, because they reflect "highly variable and uncertain linkages to events, situations, and knowledge systems" (Nespor, 1987, p.321), beliefs are unbounded in that they "always necessarily implicate the self-concept of the believer at some level, and self-concepts have wide boundaries" Abelson (1979, p.360). Such matters lead us to an examination of the function of beliefs.

On Beliefs: Their Function

Abelson (1979) and Nespor (1987) synthesise four broad characteristics of belief systems. Firstly, beliefs are derived from a person's episodic experiences. That is, "beliefs often derive their subjective power, authority, and legitimacy from particular episodes or events" that "colour or frame the comprehension of events later in time" (Nespor, 1987, p.320). In this respect, Nespor writes of a mathematics teacher, Mr. Ralston, whose undergraduate education in agriculture and many years teaching mathematics to metalworking students "led him to believe that students would be more willing to study mathematics if they could see that it had some 'practical' value" (Nespor, 1987, p.320).

Secondly, there are beliefs that an individual holds to be incontrovertibly true. From the perspective of education, this is more than just a belief in a divine being but whether a student believes academic success is due to effort or ability. Such beliefs influence greatly the effort a student is prepared to make when confronted by problems and are largely unaffected by persuasion. Thirdly, there are beliefs pertaining to alternative or ideal situations that differ from current perceptions of reality and serve to define an individual's goals. For example, Nardi and Steward's (2003) study of 13- and 14-year-old mathematics students found beliefs not only indicative of a very demotivating mathematical reality but a very strong alternative that would make their learning of mathematics more enjoyable and successful. Fourthly, there are affective or evaluative beliefs that reflect a person's response to an object that is different from the same person's knowledge about that object. For example, in an interview study of teachers' mathematics-related beliefs, Andrews (2007) found almost a third of his cohort of English teachers asserting that while they had personal curricular preferences, their statutory obligation to work within a prescribed curriculum placed it "beyond negotiation" (Andrews, 2007, p. 9).

In the following, we analyse Norwegian and Swedish upper secondary students' interview utterances for evidence of their beliefs pertaining to gjennomgang and genomgång, respectively, before discussing those beliefs against the forms and functions described above. In so doing, we acknowledge that we are trying to elicit students' testable beliefs (Abelson, 1986, p. 229), that is, beliefs "about objects within the immediate experience of the person that allow appropriate action and feedback". In inviting students to discuss the typical mathematics lesson, we are not addressing students' distal beliefs about remote objects that cannot be verified.

Methods

The data on which this paper is based derive from a comparative interview study of upper secondary students in Norway and Sweden. The aim of this study was to examine students' beliefs about the nature and purpose of school mathematics, and our interest in gjennomgang and genomgång emerged from initial analyses. In other words, data were not collected with the explicit intention of exploring students' perceptions of "going through". That being said, and acknowledging Fenstermacher's (1978, p.173) well-known assertion that what is in the mind is "accessible only by inference", one aim of the original interviews was to elicit students' perspectives on the typical lesson. In so doing, we were essentially inviting students to synthesise their many years' experience of mathematics classrooms. Such syntheses, filtered through students' experientially constructed beliefs about mathematics and its teaching, can themselves be construed as beliefs, not least because the act of synthesis both draws on and constantly recreates students' beliefs as dispositions to act (Dilworth, 2005).

To facilitate students' syntheses, group interviews, based around five broad questions, were conducted with both academic and vocational track students in both countries. Significant in this decision was the perception that group interviews facilitate exploratory research focused on "looking at a social context that is unfamiliar or new" in ways that will facilitate a better understanding of that context (Frey & Fontana, 1991, p. 177). Moreover, by:

> allowing opinions to bounce back and forth and be modified by the group, rather than being the definitive statement of a single respondent, group interviews would allow us to elaborate statements made. (Frey & Fontana, 1991, p. 178)

In Norway, the study comprised 17 interviews conducted with 42 students in three schools. Two of these schools, one in Oslo and one in Trondheim, were high-achieving academic schools, while the third was a relatively low-achieving Oslo vocational school. In Sweden the study drew on 18 interviews conducted with 50 students from four different Stockholm schools, each of which offered both academic and vocational tracks. Thus, we make no claims about the representativeness of the schools nor do we seek to generalise, although it would be reasonable to infer future lines of research.

Interviews, for which appropriate permissions had been received, were undertaken at a time convenient to the students and recorded by means of webcams on laptop computers. This decision was justified in at least four ways. Firstly, video recordings, particularly in a context where people talk over each other, enable better transcriptions than sound recordings. Secondly, video recorded interviews, which permit the capture of non-verbal communication, allow for more nuanced interpretations than sound alone. Thirdly, due to their classroom ubiquity, laptops were expected to cause less disruption than video cameras mounted on tripods. Fourthly, laptops allow the recording of data directly to their hard drives, simplifying data storage and analysis.

All interviews were transcribed and scrutinised for episodes in which gjennomgang, genomgång or their derivatives could be inferred, whether implicitly or explicitly. These episodes were then subjected to a constant comparison analysis whereby each episode was read and reread and categories of response identified and compared with each other (Boeije, 2002; Fram, 2013). With each new category, previously read episodes were reread to determine whether the new category applied to them also. This inductive process facilitates the development of theory through "categorizing, coding, delineating categories and connecting them" (Boeije, 2002, p. 393).

Results

In the following, we present our analyses of the two countries' data separately. We do this to facilitate both the reader's task and a subsequent comparative discussion. We offer the Swedish analysis first because, in our view, it allows us to present a more coherent and concise narrative.

Swedish Students' Perspectives on Genomgång

Whether academic or vocational, little variation was found with respect to Swedish students' perceptions of a genomgång's general form and function. For example, Werner summarised his perception of genomgång, with which his friend Hans agreed, as:

> He usually opens the book and opens the page that we're going to read about or the chapter, and he starts to talk about the different kinds of things that have to do with this. And he writes it all down on the white board.

Martin's independently expressed comments brought more detail to Werner's minimal description. He commented that:

> The teacher often starts with an example that maybe we pupils don't have any clue how to answer… And that's because it's a new area for us. And then he starts showing us this new method that is part of the new chapter that we're moving into… We usually listen and are free to take notes, but you don't have to. So, you listen as well as you can, I guess, and follow and try to understand... But mostly it's like a sort of, demonstration of what to do… Because after the demonstration we will be working with tasks on the topic, and we need to understand them.

Thus, a genomgång appears to be an opportunity for a teacher to demonstrate the day's new procedure. There may be occasional interactivity, but typically the role of the student is to follow the demonstration and take notes. This latter aspect was exemplified in the comments of Werner and Hans, who said:

Werner: I listen while I take notes. I write exactly what he is writing, so it memorises my brain better… (looks at Hans)
Hans: (breaks in) Yeah, I usually just listen, because I don't like writing so much... But usually I just listen and then take notes.

Throughout, students implied that genomgångs are focused on procedural aspects of mathematics, and, for most, their role is to find ways of ensuring that they understand what their teacher is showing them. There was some variation to this generally accepted view. For example, Jan noted that his:

> teacher writes an example on the board, an equation for example, and then he goes through the different rules that apply to solving the equation. And so he is trying to get everyone to understand these rules. And then also he can give us an example for the class to do together,

> and if there is someone who wants to go forward, for example, to the board, you can go and report how you do it so that the class should understand.

From Jan's description, two qualitatively different genomgångs can be inferred. The first was the procedural genomgångs described by Hans, Martin and Werner. The second, and rarely mentioned by other students, involved a public sharing of solution strategies at the board.

Occasionally, students mentioned that a genomgång was not always focused on the introduction of a new material. For example, Nadja and Ragna distinguished between two forms of genomgång:

Nadja:	He goes through a task that is going to show us what the next chapter is about and then we get one lesson for the entire chapter because the next lesson we will continue with a new one...
Interviewer:	So there's always an introduction with a task and a common activity?
Ragna:	The task from every chapter is new I guess… but he usually like next lesson goes through the harder tasks we maybe didn't do that day, that we couldn't solve, he goes through them so that we can move on to the next chapter.

For Nadja and Ragna, a genomgång may be an introduction to the next chapter, typically completed the same lesson, or it may be a demonstration of solutions to problems from the previous lesson that had proved difficult for students to complete. A genomgång conceived in this way provides a link between successive lessons.

With respect to time, students' perceptions varied. Martin believed a genomgång was typically 20 min of a lesson of one hour's duration, while Hans believed it typically lasts "fifty to fifty-five minutes and then we get five minutes for exercises". Others described something in between. For Julio it was usually "about a half-hour"; for Monika, "I think he talks for between twenty and thirty minutes", while for others it varied according to circumstances, as seen in Nadja's not atypical comment:

> It depends on how much time we need to think about things, because sometimes it takes up most part of the lesson like 30 or 35 minutes. But that's just because he really wants us to think about it… But when it's smaller tasks ... it's like 15 minutes.

In sum, despite variation in their perceptions of its length, all Swedish students seem to construe genomgång as a major element of the typical lesson that serves an important structural role in the teaching and learning of mathematics.

Norwegian Students' Perspectives on Gjennomgang

Unlike the largely consistent perspectives of the Swedish students, the extent to which gjennomgang was perceived to form an integral part of the Norwegian lesson seemed to vary according to whether students were following vocational or academic tracks. With respect to vocational students, some spoke in ways indicative of gjennomgang as a presentation of the sort of tasks that would be covered during periods of individual seatwork. For example, Matheus commented that "the teacher gives a kind of introduction to what we're going to do and then we just do exercises...", while Simen suggested that "the teacher goes through something on the board first and then … we do exercises according to what he showed on the board". Interestingly, Andreas tied such activity explicitly to his teacher's use of the textbook, commenting that:

> it was usual that we followed the chapters from start to end in the book, and for each chapter the teacher gives a sort of introduction to the chapter and then he shows on the board and then you just do the exercises.

In such comments, particularly in Simen's use of the words "goes through", it is an indication that these students were not only used to a gjennomgang-like phase of a lesson but that they believed it to be an opportunity for their teacher to introduce them to new material, albeit in unspecified ways.

Other students, however, appeared to believe that gjennomgang was a rarity. For example, Dennis commented that we "come to class, get out our books, get our exercises on the board… and then we do them", while John-Martin observed that our:

> teacher tells us what we should do, tells us which exercises and stuff we have to do and then we do that… there has been a lot of exercise on the PC lately.

Neither Dennis nor John-Martin saw these introductions as anything but a specification of what exercises they were expected to undertake. There was no indication in either of their utterances of explanation or demonstration. This sense of teachers rarely going through new material was highlighted in a brief exchange between Ahmadi and Mukhtar. They said:

Ahmadi: Yes it starts as a normal lesson, a bit of noise and then…
Mukhtar: (Finishing Ahmadi's sentence) We get different exercises every day and we work with those.

Moreover, confirming the rarity of gjennomgang for some students, Viktor commented that "we go into the classroom, take a PC and start working". In other words, at least as far as Viktor believed, teachers do not engage with the whole class but devolve teaching to some form of computer-based material.

Academic track students' perceptions were not dissimilar to those of the first group of vocational students. Amelie, commented that:

> there is often a kind of going through first, with new information that we kinda have not learnt about before… or a bit of repetition, it depends... at the start of the week there is often a lot at first, and then we repeat later in the week.

Embedded in Amelie's comments is a perception that gjennomgang serves to either introduce new material or revisit old. However, her comments offer little by way of detail with respect to how a gjennomgang typically plays out. In this respect it is important to note that while almost all academic track students spoke about their teacher "going through" the material that would be covered during the lesson, only a few offered comments from which details could be inferred. In this respect, Mats observed that:

> usually the teacher comes in and if we are starting a new topic he will go through that topic, er... methods for doing exercises.

In addition, Emily noted that her typical lesson:

> starts with the teacher explaining what we'll go through that lesson… we go through something new every lesson more or less.

While Sarah noted that "it starts with the teacher being at the board and going through formulas and... example exercises… and then we do exercises". In such comments lies a common belief concerning teachers "going though" of new material; in their "explaining", they show "methods for doing exercises" and undertake "example exercises". As such the typical gjennomgang appears procedurally focused and highly teacher-centred.

Other students indicated that their teachers' gjennomgangs created extensive notes. For example, Emelie mentioned that "there is always a lot of notes on the board, and examples are used a lot", while Pauline recalled that her teacher "writes a lot on the board, first he says which chapter it is and he has some example exercises". Other students indicated that gjennomgangs are not the brief episodes implied by the vocational students above but often lasted much of the lesson. For example, Petter spoke of how his teacher:

> often goes through an example on the board... when we have been through some new material... we don't do too many exercises I feel, it is not like he opens the book and tells us what exercises to do.. a lot of the lesson is spent explaining the maths... and that takes up a lot of the time.

In similar vein, Øyvind commented not only that "there is often a bit of talk first" but also that this "talk can drag on a bit, but that's the way it is".

Discussion

Our aim in this paper, drawing on group interviews undertaken with no such purpose originally in mind, was to investigate Norwegian and Swedish upper secondary students' beliefs about the nature and purpose of gjennomgang and genomgång. In so doing, we set out, also, to consider the extent to which their belief interview utterances resonated with earlier literature on the form and function of beliefs, and it is here that we begin our discussion.

Both Abelson (1979) and Nespor (1987) write of beliefs as non-consensual and unbounded. The data yielded by the interviews indicate that while there may have been elements of consensus seeking within the interview groups, as in the interchanges above between Werner and Hans or Ahmadi and Mukhta, there were no indications that students attempted to legitimate their perspectives by reference to external sources. Also, the variation in their utterances, even when their foci were similar, confirmed their unboundedness. Thus, we would argue that from the perspective of their form, students' utterances can be unproblematically construed as beliefs.

From the perspective of function, students' utterances were unproblematically derived from those episodic experiences that ultimately "colour or frame the comprehension of events later in time" (Nespor, 1987, p.320). Their comments reflected not random thoughts but events, both recent and past, as seen in Nadja's comments about the varying lengths of genomgång she and Ragna had experienced. With respect to notions of certainty, many students' utterances drew on words like "usually", "typically" and "mostly", indicating that while things may not always be the same, there is a routine predictability in their interpretation of genomgång or gjennomgang. This predictability, or experiential truth, was further supported by Emelie's, "there is always a lot of notes on the board", and Werner's, "I write exactly what he is writing". From John-Martin's observation that "there has been a lot of exercises on the PC lately" can be inferred an awareness of an alternative comprising a reality less monotonous and more inspiring than the current and, finally, from Øyvind's comment that his teacher's "talk can drag on a bit, but that's the way it is", can be inferred an evaluation. In sum, our view is that these students' utterances, which were enhanced and refined by the use of group interviews (Frey & Fontana, 1991), yielded sufficient evidence for them to be construed as indicators of their beliefs about the form and function of gjennomgang and genomgång.

So, with respect to the particular aims of the analyses, what has been learned about gjennomgang and genomgång? Do they refer to a commonly understood practice? Are they didactical folkways? To both questions, we offer a tentative, yes, although it is important to acknowledge subtle cross-national variations. For all Swedish, Norwegian academic and some Norwegian vocational students and gjennomgang and genomgång referred to a lesson episode where teachers "go through" something, a process that seems to take two forms. During the first, which we have labelled as an *instructional* gjennomgang or genomgång, teachers demonstrate or explain new procedures. Initially we were minded to describe these as explanatory, but didactical explanations entail the explicit collaboration of all participants (Leinhardt, 2001), which was absent from students' descriptions. Thus, we see *instructional* as a more accurate description. The second we describe as a *problem-solving* gjennomgang or genomgång, where teachers demonstrate solutions to challenging problems from previous lessons. Interestingly, Jan, a Swedish academic track student, alluded to a genomgång that involved a public sharing of solution strategies in the manner described by Andrews (2009a). However, this was mentioned only in his group's interview in relation to a teacher, a relatively recent immigrant to Sweden, who had experienced his professional training outside the Scandinavian countries, and could not, therefore, have assimilated the folkways of his Swedish colleagues. Finally, with regard to similarities, gjennomgang and genomgång typically take up long periods of the lesson and involve extensive use of the board and an expectation, albeit implicit, that students make notes.

The most significant variation emerged from the utterances of some of the Norwegian vocational students for whom gjennomgang seemed a rarity. These students, implying that teaching was being delegated to a book or computer, described a fleeting and unspecified introduction to a new material. That is, students seemed to be describing a *directive* gjennomgang that lasts just as long as it takes for teachers to explain the lesson's activities.

Finally, unlike the Norwegian vocational interviews, no Swedish interviews were undertaken in the socially deprived suburbs of Stockholm. Had such interviews taken place, it remains a conjecture whether the *directive* gjennomgang seen in Oslo would have been reflected in Stockholm. As it is, the limited evidence of this study indicates that gjennomgang and genomgång are construed similarly in the two countries, albeit with evidence of some local variation.

References

Abelson, R. (1979). Differences between belief and knowledge systems. *Cognitive Science, 3*(4), 355–366. https://doi.org/10.1207/s15516709cog0304_4

Abelson, R. (1986). Beliefs are like possessions. *Journal for the Theory of Social Behaviour, 16*(3), 223–250. https://doi.org/10.1111/j.1468-5914.1986.tb00078.x

Andrews, P. (2007). The curricular importance of mathematics: A comparison of English and Hungarian teachers' espoused beliefs. *Journal of Curriculum Studies, 39*(3), 317–338. https://doi.org/10.1080/00220270600773082

Andrews, P. (2009a). Mathematics teachers' didactic strategies: Examining the comparative potential of low inference generic descriptors. *Comparative Education Review, 53*(4), 559–581. https://doi.org/10.1086/603583

Andrews, P. (2009b). Comparative studies of mathematics teachers' observable learning objectives: Validating low inference codes. *Educational Studies in Mathematics, 71*(2), 97–122. https://doi.org/10.1007/s10649-008-9165-x

Andrews, P., & Sayers, J. (2013). Comparative studies of mathematics teaching: Does the means of analysis determine the outcome? *ZDM, 45*(1), 133–144. https://doi.org/10.1007/s11858-012-0481-3

Boeije, H. (2002). A purposeful approach to the constant comparative method in the analysis of qualitative interviews. *Quality and Quantity, 36*(4), 391–409. https://doi.org/10.1023/a:1020909529486

Brown, C., & Cooney, T. (1982). Research on teacher education: A philosophical orientation. *Journal of Research and Development in Education, 15*(4), 13–18.

Buchmann, M. (1987). Teaching knowledge: The lights that teachers live by. *Oxford Review of Education, 13*(2), 151–164. https://doi.org/10.1080/0305498870130203

Dilworth, J. (2005). The reflexive theory of perception. *Behavior and Philosophy, 33*, 17–40.

Ernest, P. (1989). The knowledge, beliefs and attitudes of the mathematics teacher: A model. *Journal of Education for Teaching, 15*(1), 13–33. https://doi.org/10.1080/0260747890150102

Fenstermacher, G. D. (1978). A philosophical consideration of recent research on teacher effectiveness. *Review of Research in Education, 6*(1), 157–185. https://doi.org/10.3102/0091732x006001157.

Fram, S. M. (2013). The constant comparative analysis method outside of grounded theory. *The Qualitative Report, 18*(1), 1–25. url: http://nsuworks.nova.edu/tqr/vol18/iss1/1

Frey, J. H., & Fontana, A. (1991). The group interview in social research. *The Social Science Journal, 28*(2), 175–187. https://doi.org/10.1016/0362-3319(91)90003-M

Green, T. F. (1971). *The activities of teaching*. London: McGraw Hill.

Häggström, J. (2006). The introduction of new content: What is possible to learn? In D. Clarke, J. Emanuelsson, E. Jablonka, & I. Mok (Eds.), *Making connections: Comparing mathematics classrooms around the world* (pp. 185–199). Rotterdam: Sense Publishers.

Knipping, C. (2003). Learning from comparing: A review and reflection on qualitative oriented comparisons of teaching and learning mathematics in different countries. *ZDM, 35*(6), 282–293. https://doi.org/10.1007/bf02656692

Leinhardt, G. (2001). Instructional explanations: A commonplace for teaching and location for contrast. In V. Richardson (Ed.), *Handbook of research on teaching* (pp. 333–357). Washington: American Educational Research Association.

Nardi, E., & Steward, S. (2003). Is mathematics T.I.R.E.D? A profile of quiet disaffection in the secondary mathematics classroom. *British Educational Research Journal, 29*(3), 345–367. https://doi.org/10.1080/01411920301852

Nespor, J. (1987). The role of beliefs in the practice of teaching. *Journal of Curriculum Studies, 19*(4), 317–328. https://doi.org/10.1080/0022027870190403

Op 't Eynde, P., De Corte, E., & Verschaffel, L. (2006). Epistemic dimensions of students' mathematics-related belief systems. *International Journal of Educational Research, 45*(1–2), 57–70. https://doi.org/10.1016/j.ijer.2006.08.004

Skott, J. (2009). Contextualising the notion of 'belief enactment'. *Journal of Mathematics Teacher Education, 12*(1), 27–46. https://doi.org/10.1007/s10857-008-9093-9

Stigler, J. W., & Hiebert, J. (1999). *The teaching gap*. New York: The Free Press.

Temporal Norms of the Typical Mathematics Lesson: Norwegian and Swedish Students' Perspectives

Mona Nosrati and Paul Andrews

Abstract In this paper, we present a number of Swedish and Norwegian high-school students' descriptions of a 'typical mathematics lesson'. These descriptions are subsequently considered in light of Foucault's discussion of timetables and discipline originally employed in armies and subsequently in schools throughout Europe. We contend that the structure of the typical mathematics lesson – though arguably unremarkable in and of itself – is both culturally normative and historically situated. Thus, in its very simplicity, it offers a window on the temporal norms imposed on and maintained in educational institutions in general and in the mathematics subject in particular.

Keywords Foucault • Mathematics • Cultural norms • Temporal norms • Educational institutions

Introduction

Several studies of the 1990s, particularly the Trends in International Mathematics and Science Study (TIMSS) video studies (Hiebert et al. 2003; Stigler, Gonzales, Kawanaka, Knoll, & Serrano, 1999) and the Survey of Mathematics and Science Opportunities (Schmidt et al. 1996), have concluded that mathematics teaching, drawing on a subconscious routine and consistent re-enactment of particular pedagogies, is culturally normative. That is, teachers of mathematics adhere, consciously or otherwise, to a culturally determined script (Andrews & Sayers, 2013).

M. Nosrati (✉) • P. Andrews
Matematikksenteret, Norwegian University of Science and Technology and Stockholm University, Stockholm, Sweden
e-mail: mona.nosrati@matematikksenteret.no

H. Palmér, J. Skott (eds.), *Students' and Teachers' Values, Attitudes, Feelings and Beliefs in Mathematics Classrooms*,
https://doi.org/10.1007/978-3-319-70244-5_12

Motivations for investigating this script have frequently stemmed from the desire to understand the causes of the consistently high performance of East Asian students, when compared with their Western counterparts. However, these and many other studies have examined the typical lesson from an observer's perspective. Few have considered the student view. However, as Bourdieu (1990) notes, by postulating an objective model of any practice through outside observation only, 'the analyst reduces the agents to the status of automata or inert bodies moved by obscure mechanisms towards ends of which they are unaware' (p. 98).

To avoid doing so when seeking to understand educational practices, the student perspective must be considered. Learning what students see their typical mathematics lesson to be about is important, because it forms the basis of their experience with the subject and hence plays a central part in the development of their mathematics-related beliefs. These beliefs in turn play a significant role in determining students' engagement with and subsequent learning of the subject (Hannula 2006; Leder and Forgasz, 2002; Ma and Kishor 1997).

With respect to our understanding of students' beliefs about the nature of a typical lesson, little research has been undertaken with such an explicit focus. A small number of qualitative studies in the UK and the USA (e.g. Boaler, 1998; Nardi & Steward, 2003; Ryan & Patrick, 2001; Turner et al., 2002) have examined the impact of classroom practices on students' mathematics-related affect, and in so doing, they have alluded to the typical lesson experienced by students. In a similar vein, a number of quantitative studies have implicitly addressed the typical lesson by investigating various aspects of student affect in relation to classroom practices derived from the literature (e.g. Fall & Roberts, 2012; Rakoczy et al., 2013). Our aim in this paper is to make a more explicit contribution to the field.

In the following, we present and discuss a number of Swedish and Norwegian high-school students' descriptions of a 'typical mathematics lesson', and consider how this typicality might be captured by a set of 'temporal norms' described by Foucault (1977) in his discussion of timetables and discipline originally employed in armies and subsequently in schools throughout Europe.

Method

We have conducted semi-structured interviews with 47 Swedish and 41 Norwegian high-school students (aged 16–18) in both countries selecting about half the students from vocational and half from academic tracks. In Sweden, the interviews were conducted at four different Stockholm schools, each of which offered both academic and vocational tracks. In Norway the academic track interviews were conducted at two different schools, one in Oslo and one in Trondheim. Both schools have high entrance requirements and are considered to be among the best high schools in their respective cities. The vocational track interviews were conducted with students from a high school in Oslo which could be said to lie at the very opposite end of the spectrum, with low entrance requirements and a poor academic reputation.

Students were interviewed in pairs or threes so that they would have the opportunity to discuss among themselves rather than just with the interviewer, and the interviews were structured around the following four questions (with follow-up questions where appropriate):

1. How would you describe a typical mathematics lesson at school?
2. What do you think is the purpose of compulsory school mathematics?
3. What do you think mathematics as a subject has to offer to those who engage with it?
4. If you could say something about the nature of mathematics education to those in charge of the educational system, what would it be?

The students were asked to describe both their current experiences in high school and their experiences throughout earlier years of compulsory schooling. Video and sound were recorded on a laptop web camera, and all interviews were transcribed. The student responses were then subjected to a constant comparison analysis (Glaser, 1965) whereby each episode was read and reread and categories of response identified and compared with each other. With each new category, previously read episodes were reread to determine whether the new category applied to them also.

The Typical Lesson: Why Care?

The question about the typical lesson in many ways sets the scene for the interviews and for the questions that followed. The question does not at first seem to ask for much in terms of beliefs, opinions or knowledge about mathematics. It simply asks for a description, and as far as questions go, it could be said to border on the trivial. Indeed one might argue that the truly interesting data would be the students' responses to questions 2 and 3 above, concerning young people's conceptions of the purpose of this subject that they are made to endure for at least 10 years.

However, in eliciting students' perspectives on the typical lesson, we were essentially inviting them to synthesise their many years' experience of mathematics classrooms. Such syntheses, filtered through students' experientially constructed beliefs about mathematics and its teaching, can themselves be construed as beliefs. Firstly, it is because each synthesis reflects one student's beliefs as a perception of reality (Van den Bossche, Gijselaers, Segers, & Kirschner, 2006), and secondly, the act of synthesis both draws on and constantly recreates students' dispositional beliefs (Dilworth, 2005). Thus it could be argued that any beliefs expressed in response to questions 2 and 3 are a direct consequence of what takes place in the typical lesson, and as such the student responses to question 1 deserve some careful attention.

The Typical Lesson: Student Descriptions

Norwegian and Swedish Academic Track Students

The Norwegian and Swedish academic track students' accounts of the typical mathematics lesson were remarkably similar and highly consistent across interviews in both countries. Mårten, a Swedish student in the first year of the natural science and art programme, gave the following description:

> Well, the structure is that first the teacher talks about the chapter we are moving into. And then we work by ourselves mostly from the books, doing different exercises and so on… That's pretty much it. And so we work through… we follow the book, like chapter after chapter. Yeah, the teacher often has a genomgång (going-through) and so on about what we're doing.

Although the wording and detail differed, other Swedish responses from the different schools conveyed the same overall message. For example, Frans commented that:

> Normally we go through what we will be doing; first we have a short summary of what we will do, then the teacher usually holds a genomgång (going-through) on the section we are going to work with… And then we usually undertake exercises on what we have learned.

Norwegian academic track students from Oslo also had a similar story to tell, as seen in Mari's comment that 'usually the teacher comes in and if we are starting a new topic he will go through that topic emm... methods for doing exercises, eh and then we do exercises. That's the average…', or Emily's belief that it 'starts with the teacher explaining what we'll go through that lesson.. we go through something new every lesson more or less.. then you carry on with exercises'. It is worth noting that in the case of a double lesson – which is another way of saying that there is more *time* – it appears that the lesson structure is simply repeated twice, as confirmed by Mina, who commented that 'we usually go through a subchapter in plenary and then do exercises for it.. if we have a double lesson we often go through one more subchapter too… and do exercises for that'.

Remarkably (or perhaps not?) the answers did not differ much in interviews with students around 500 km away in Trondheim, as found in the interview with Line and Malin:

Line: It is typical that we start with a bit of teaching on the board, what we will carry on with…

Malin: yes, what we have done, and what we will be doing.. a bit of information like that, and then maybe we go through a new subchapter or topic, and then we work with exercises…

Øyvind, another Trondheim interviewee, noted that there often is a bit of talk first and that 'that talk can drag on a bit ... but it is mostly working on exercises'.

In sum, and acknowledging some variations in the descriptions, the typical lesson was described as consisting of two main parts: (1) The teacher goes through something on the board and, in doing so, defines the topic for the lesson. This is usually based on a chapter or subchapter from a textbook, and examples of how to solve specific types of exercises are given. (2) Students do corresponding exercises from the textbook.

It is also worth noting that the exercises were largely (though not exclusively) reported to be done individually. The following response from two Norwegian students captures what appeared to be a common experience among interviewees from both countries:

Interviewer: do you tend to work individually or in groups?
Kristine: we work very little in groups..
Emily: very little in groups..
Kristine: you could potentially whisper to the person sitting next to you.. ask 'did you get that?' and 'did you get that too on that exercise?'.. but.. yeh..

Norwegian and Swedish Vocational Track Students

The vocational track students' responses also referred to a two-part lesson in which the first part consisted of the teacher *going through* something on the board followed by largely individual work on exercises – as described for example by Björn, a Swedish student:

Björn: We normally listen to the teacher when he goes through (gå igenom) the next chapter and then work on the chapter after he has finished talking about it
Interviewer: What does it mean to work with the chapter?
Björn: Doing maths exercises from the book
Interviewer: OK, when you work from the book can you work with someone else?
Björn: You can work with someone else but usually you do it on your own.

The same structure was once again also described by the Norwegian students. For example, Andreas commented that:

> it is usual that we follow the chapters from start to end in the book, and for each chapter the teacher gives a sort of introduction to the chapter and then he shows on the board and then you just do the exercises..

While Simen believed:

> it is just that the teacher goes through something on the board first, and then he puts up some exercises and then we do those exercises according to what he showed on the board.. and then he goes around and helps those who need help.. really it is a pretty simple lesson

The main *difference* between the accounts given by academic and vocational track students was seen in the Norwegian data, where the latter group in several interviews referred to high levels of classroom noise. In this respect, Ahmadi observed that 'yes it starts as a normal lesson, a bit of noise and then…', while Ali, speaking in a different interview, commented that:

> because of others, I can't concentrate properly. There is a lot of chaos and they just sit there and make noise.. and they threaten us like 'give us the exercises or you'll get beaten outside' and stuff.. that's on exercises we have to hand in that is

In yet another interview, Robin noted that 'there are so many making noise all the time', and Asta nodded in agreement and added 'yes, very hard to get peace and quiet'.

None of the Swedish students or Norwegian academic track students interviewed made any reference to noise. This finding is perhaps not so surprising given the nature and status of the Norwegian vocational track school chosen for interview purposes. However, it is certainly remarkable that despite this 'noise', the typical structure of the lesson appeared to be maintained where possible. And although Swedish vocational track students did not report noise as an issue, it was clear that the lesson structure did not necessarily secure their full attention, as noted by André:

> I listened to him quite a lot and I answered a lot of the questions he asked but (…) I actually didn't try that much because I slept a lot in the lessons (...) and then we might just start working from the book… You could go outside and sit if you wanted a more quiet place. So that was very nice. But we worked a lot from the books.

Finally, it is also worth noting that whereas the start of a lesson was very clearly defined and described by the interviewees (of both vocational and academic tracks), the end of a lesson was much less so. There was no concrete mention in the data of what takes place at the end of each lesson, and in the vocational track interviews, there was even an indication that once you have done your exercises you can just leave the classroom. As Matheus explained:

> the teacher gives a kinda introduction to what we're going to do, and really then we just do exercises.. that is it.. and when we are done we the exercises we can just go out.

Discussion

Why write a paper about the typical lesson? What will be added to our knowledge by doing so? It has been said that writing an academic paper is all about stating the obvious with an air of surprise. Here even the air of surprise is arguably hard to produce, at least for those who are familiar with the educational systems in question. There is nothing remarkable about the responses described above. Upon reflection what is striking is the *absence* of variation. An absence of a single student out of 89 saying 'well, that is a very difficult question to answer, because it really varies so much from lesson to lesson'.

Hence the purpose of looking at the structure of the typical lesson must be less about informing the reader about *what* this structure is and more about bringing this absence of variation into our collective consciousness by considering *why* this structure exists and *where* it might have come from (where something is culturally normative, it is also historically rooted).

For to state that mathematics lessons are conducted repeatedly and consistently in the same way across classrooms, schools and even countries is one thing. Another is to acknowledge what this means in terms of options excluded. There is an infinite number of ways (or let's just say a large number of ways for those who find the infinite daunting) in which 45–90 min could be spent within the world of mathematics, and yet we do, repeatedly, consistently, across classrooms, schools and countries pick just one! If it were merely a game of chance, this would have been less likely than winning the lottery. But of course, the engrainedness and widespreadness of the lesson structure have not come about by chance (and nor does it seem like anyone is celebrating for that matter). Rather, the practice has emerged through history – for reasons we may have forgotten or never been aware of in the first place – and with time it has been reinforced and embedded in the culture and thus allowed to live a quiet life in the subconscious of a population without ever being brought out in the light to be studied with curious eyes.

But what if we try to do just that? What is it really that we see? We see that *time* is carefully divided and its use highly regulated. The time that students spend at school is – at most educational institutions – carefully divided according to a timetable, and a mathematics 'lesson' is a clearly delineated period of time in that timetable during which mathematics is to be undertaken. How the mathematics lesson itself is structured is a further subdivision of time, with an introductory part of 'going through' and a 'working on exercises' part. The 'going through' part closely follows a textbook and effectively dictates how the second part of the lesson is to be spent. Aside from being restricted in terms of mathematical topic, this second part is itself divided – in an ordered sequence – between a number of prescribed exercises, and even the very steps to be taken in order to solve these exercises are presented in the form of an example at the beginning of the lesson. Thus even the time spent on each exercise is effectively divided and regulated. In other words, the structure of the typical mathematics lesson could be seen as a mere continuation of the general timetable, on an increasingly smaller and more detailed scale.

But where does this detailed form of timetabling come from? Foucault (1977) argues that throughout Europe, the organisation of a strict timetable historically emerged from the army and was subsequently adopted by schools. Notably:

> The principle that underlay the time-table in its traditional form was essentially negative; it was the principle of non-idleness: it was forbidden to waste time, which was counted by God and paid for by men; the time-table was to eliminate the danger of wasting it - a moral offence and economic dishonesty. (p 154)

Paired with an ideology of discipline, the institutions could make the very most of any available time, because:

> Discipline ... poses the principle of a theoretically ever-growing use of time: exhaustion rather than use; it is a question of extracting, from time, ever more available moments and, from each moment, ever more useful forces. This means that one must seek to intensify the use of the slightest moment, as if time, in its very fragmentation, were inexhaustible or as if, at least by an ever more detailed internal arrangement, one could tend towards an ideal point at which one maintained maximum speed and maximum efficiency. (ibid, p. 154)

Foucault further notes that the Prussian army regulations of 1743 (which the rest of Europe imitated) laid down six stages to bring the weapon to one's foot, four to extend it, 13 to raise it to the shoulder and so on, and that schools, also arranged to intensify the use of time, followed analogous practices by carefully regulating the operations performed by pupils under the direction of monitors and assistants:

> ... so that each passing moment was filled with many different, but ordered activities; and ... orders imposed on everyone temporal norms that were intended both to accelerate the process of learning and to teach speed as virtue.

Scandinavian schools (or educational systems) today would hardly subscribe to such a military view of learning focused on timetables, discipline and imposed temporal norms. Quite on the contrary we find, for example, that the Norwegian curriculum aims for mathematics clearly state that:

> The learning shifts between enquiry-based, playful, creative and problem solving activities and the practice of skills ... Provisions must be made such that both girls and boys have rich experiences with the mathematics subject, creating positive attitudes and a solid subject competence. This forms the basis for lifelong learning. (Læreplan i Matematikk, our translation)

There is an emphasis here on *enquiry*, *playfulness*, *creativity* and *rich* experiences, which resonate strongly with many years of educational research recommendations (e.g. Cuoco, Goldenberg, & Mark, 1996, Goos, 2004; Stein, Engle, Smith, & Hughes, 2008; Wæge 2007). And yet, our data clearly indicate that the imposition of temporal norms is inadvertently maintained and reproduced through the typical lesson structure, even where both teachers and students are seemingly squirming under its restrictions. We see here a temporal practice in which it could be argued that 'the Dead seize the Living' (Bourdieu, 1980). That is, the dead institutions and conventions wrought by the movement of history grab hold of, inscribe themselves into and seize the practices of the living. Or as argued by Marx (1852):

> Men make their own history, but they do not make it as they please; they do not make it under self-selected circumstances, but under circumstances existing already, given and transmitted from the past. The tradition of all dead generations weighs like a nightmare on the brains of the living.

Even so, one might come to wonder how the temporal norms of dead generations of educational systems can be maintained if educational ideologies have changed – as curriculum aims and research recommendations suggest they have. Perhaps the answer is to be found in Paolo Freire's contention that it is not education which somehow moulds society but rather society which, according to its particular structure, shapes education in relation to the ends and interests of those who control the power in that society. Then, for as long as the primary interests of those who control power in society are in international test scores and differentiation of students (e.g. Sjøberg, 2014), the breaking down of the mathematics curriculum to ever-increasing and well-defined competence aims that can be tested in exams is an inevitable consequence. Such a fragmentation of learning goals in turn leads to a corresponding fragmentation of time akin to that which emerged from assuming army discipline to also be an educational virtue. In both cases, to borrow Foucault's words one final time:

> the 'seriation' of successive activities makes possible a whole investment of duration by power: the possibility of a detailed control and a regular intervention (of differentiation, correction, punishment, elimination) in each moment of time; the possibility of characterizing, and therefore of using individuals according to the level in the series that they are moving through; ... Power is articulated directly onto time; it assures its control and guarantees its use. (Foucault, 1977, p. 160)

Conclusion

We argued initially that learning what students see their typical mathematics lesson to be about is important, because it forms the basis of their experience with the subject and hence plays a central part in the development of their mathematics-related beliefs. Through interviews with high-school students, we have found that – from their perspective – there does indeed seem to be such a thing as a typical lesson, and this typicality appears to derive from a set of relatively stringent temporal norms reminiscent of times where discipline and timetables were thought to form the pillars of education.

In other words, despite the great emphasis in recent research and curricular aims on the importance of enquiry-based learning and creativity – which require relatively large degrees of temporal freedom (Nosrati, 2015) – we find instead that mathematics lessons bear clear marks of temporal control and restriction. The reasons for this may be historical, cultural and ultimately practical, and one could not expect that the publication of academic papers on the issue should have lead to major changes in the 50 or so years since mathematics education emerged as a discipline in its own right. However – and crucially – this does not mean that thinking and talking about alternative ways of doing things (with students as well as with teachers, researchers or politicians) should be considered a futile exercise. In this we align ourselves with Bourdieu's (1990) view that even 'the simple possibility that things might proceed otherwise is sufficient to change the experience of practice and, by the same token, its logic' (p. 99). The very uncertainty that arises from encouraging alternative strategies allows for a reconsideration of temporal norms, with the potential to free time (at least temporarily) from the minutest of timetables: 'To reintroduce uncertainty is to reintroduce time' (ibid.).

References

Andrews, P., & Sayers, J. (2013). Comparative studies of mathematics teaching: Does the means of analysis determine the outcome? *ZDM, 45*(1), 133–144. https://doi.org/10.1007/s11858-012-0481-3

Boaler, J. (1998). Open and closed mathematics: Student experiences and understandings. *Journal for Research in Mathematics Education, 29*(1), 41–62. https://doi.org/10.2307/749717

Bourdieu, P. (1980). Le mort saisit le vif [Les relations entre l'histoire réifiée et l'histoire incorporée]. *Actes de la recherche en sciences sociales, 32*(1), 3–14.

Bourdieu, P. (1990). *The logic of practice*. Stanford, CA: Stanford University Press.

Cuoco, A., Goldenberg, E., & Mark, J. (1996). Habits of mind: An organizing principle for mathematics curricula. *Journal of Mathematical Behavior, 15*(4), 375–402.

Dilworth, J. (2005). The reflexive theory of perception. *Behavior and Philosophy, 33*, 17–40.

Fall, A.-M., & Roberts, G. (2012). High school dropouts: Interactions between social context, self-perceptions, school engagement, and student dropout. *Journal of Adolescence, 35*(4), 787–798. https://doi.org/10.1016/j.adolescence.2011.11.004

Formål, læreplan i matematikk. Retrieved from: http://www.udir.no/kl06/mat1-04/Hele/Formaal/?lplang=nob&read=1

Foucault, M. (1977). *Discipline and punish: The birth of the prison*. New York: Vintage.

Glaser, B. (1965). The constant comparative method of qualitative analysis. *Social Problems, 12*(4), 436–445.

Goos, M. (2004). Learning mathematics in a classroom community of inquiry. *Journal for Research in Mathematics Education, 35*(4), 258–291. https://doi.org/10.2307/30034810

Hannula, M. (2006). Motivation in mathematics: Goals reflected in emotions. *Educational Studies in Mathematics, 63*(2), 165–178. https://doi.org/10.1007/s10649-005-9019-8

Hiebert, J., Gallimore, R., Garnier, H., Bogard Givvin, K., Hollingsworth, H., Jacobs, J., et al. (2003). *Teaching mathematics in seven countries: Results from the TIMSS 1999 video study*. Washington, DC: National Center for Educational Statistics.

Leder, G., & Forgasz, H. (2002). Measuring mathematical beliefs and their impact on the learning of mathematics: A new approach. In G. Leder, E. Pehkonen, & G. Törner (Eds.), *Beliefs: A hidden variable in mathematics education* (pp. 95–113). Dordrecht, The Netherlands: Kluwer.

Ma, X., & Kishor, N. (1997). Assessing the relationship between attitude toward mathematics and achievement in mathematics: A meta-analysis. *Journal for Research in Mathematics Education, 28*(1), 26–47. https://doi.org/10.2307/749662

Marx, K. (1852). The Eighteenth Brumaire of Louis Bonaparte, original version published in the first number of the monthly. *Die Revolution*.

Nardi, E., & Steward, S. (2003). Is mathematics T.I.R.E.D.? A profile of quiet disaffection in the secondary mathematics classroom. *British Educational Research Journal, 29*(3), 345–367. https://doi.org/10.1007/bf02656692

Nosrati, M. (2015). Temporal freedom in mathematical thought: A philosophical–empirical enquiry. *The Journal of Mathematical Behavior, 37*, 18–35.

Rakoczy, K., Harks, B., Klieme, E., Blum, W., & Hochweber, J. (2013). Written feedback in mathematics: Mediated by students' perception, moderated by goal orientation. *Learning and Instruction, 27*, 63–73. https://doi.org/10.1016/j.learninstruc.2013.03.002

Ryan, A. M., & Patrick, H. (2001). The classroom social environment and changes in adolescents' motivation and engagement during middle school. *American Educational Research Journal, 38*(2), 437–460. https://doi.org/10.3102/00028312038002437

Schmidt, W. H., Jorde, D., Cogan, L. S., Barrier, E., Gonzalo, I., Moser, U., et al. (1996). *Characterizing pedagogical flow: An investigation of mathematics and science teaching in six countries*. Dordrecht, The Netherlands: Kluwer.

Sjøberg, S. (2014). Pisa-Syndromet - Hvordan norsk skolepolitikk blir styrt av oecd. *Nytt Norsk Tidsskrift, 31*, 30–43.

Stein, M. K., Engle, R. A., Smith, M. S., & Hughes, E. K. (2008). Orchestrating productive mathematical discussions: Five practices for helping teachers move beyond show and tell. *Mathematical Thinking and Learning, 10*, 313–340. https://doi.org/10.1080/10986060802229675

Stigler, J., Gonzales, P., Kawanaka, T., Knoll, S., & Serrano, A. (1999). *The TIMSS videotape classroom study*. Washington, DC: National Center for Educational Statistics.

Turner, J. C., Midgley, C., Meyer, D. K., Gheen, M., Anderman, E. M., Kang, Y., et al. (2002). The classroom environment and students' reports of avoidance strategies in mathematics: A multimethod study. *Journal of Educational Psychology, 94*(1), 88–106.

Van den Bossche, P., Gijselaers, W. H., Segers, M., & Kirschner, P. A. (2006). Social and cognitive factors driving teamwork in collaborative learning environments: Team learning beliefs and behaviors. *Small Group Research, 37*(5), 490–521. https://doi.org/10.1177/1046496406292938

Wæge, K. (2007). *Elevenes motivasjon for å lære matematikk og undersøkende matematikkundervisning*. (PhD), Norwegian University of Science and Technology, Trondheim.

Grade 9 Students' Reasoning About Division of Fractions: What Are their Arguments Anchored in?

Lovisa Sumpter

Abstract This paper studies secondary school student's mathematical reasoning when solving tasks about fractions. The aim is to explore what the mathematical foundation is replaced with in their reasoning when reasoning is classified as imitative. Two different foundations were found: incorrect mathematical properties not relevant to the task and beliefs about mathematics and mathematics education. The results suggest that a focus on reasoning provides additional information about students' knowledge about fractions beyond standard error analysis.

Keywords Arithmetics • Beliefs • Fractions • Mathematical reasoning • Secondary level

Introduction

One of the challenges for students in school mathematics is to understand rational numbers (Nunes & Bryant, 2009). A possible explanation is the complexity of this concept. You can, for instance, see fractions as a measurement (e.g. half a pie) or as an arithmetic operation, division (Marshall, 1993). Nunes and Bryant (2009) concluded in their review of research about fractions that it is crucial for children to learn it both as a quantity and as a division. Also, their review shows that the relationship between these two views, quantity and division, doesn't come automatically, meaning teaching about fractions also needs to encompass relationship between different representations. A student's view of fractions affects their

L. Sumpter (✉)
Department of Mathematics and Science Education, Stockholm University, Stockholm, Sweden
e-mail: lovisa.sumpter@mnd.se.se

H. Palmér, J. Skott (eds.), *Students' and Teachers' Values, Attitudes, Feelings and Beliefs in Mathematics Classrooms*,
https://doi.org/10.1007/978-3-319-70244-5_13

reasoning: children are more successful ordering fractions in magnitude in situations that involve division than seeing fraction as a quantity (Nunes & Bryant, 2009). Hence, how you construct your reasoning and on which mathematical properties, this including relationships, determines how successful you are.

One of the reasons to focus on understanding of fractions is because it is an important step towards algebra (Fenell et al., 2008; Norton & Hackenberg, 2010). Confusion with fractions will thus make development of algebra more difficult. This paper aims to study student's arguments when solving tasks about fractions. The research questions are as follows: (1) When imitative reasoning, what is replacing the mathematical foundation? (2) What beliefs about mathematics and mathematics education are indicated in students' arguments?

Background

The starting point for this paper is that rational numbers are numbers in the domain of quotients (Brousseau, Brousseau, & Warfield, 2007), meaning numbers that could be defined as *a*/*b*. Children's different way of learning various aspects of rational numbers has been in focus for several research studies (Nunes & Bryant, 2009). As mentioned above, a plausible explanation of the problems with learning fractions might be due to the number of different interpretations including representations. Fractions can be interpreted as part of a whole, ratio, measure, relation and operator (Kieren, 1976, 1993; Marshall, 1993). A lot of the previous research stress the importance of learning about fractions from a multidimensional perspective (Nunes & Bryant, 2009) but also the relations between these interpretations (Nunes & Bryant, 2009). Most of the errors students make when solving tasks about fractions, such as the errors presented in Padberg (1989 in Engström, 1997), stem from confusions students experience when constructing understanding about concepts of numbers including fractions (Siegler & Lortie-Forues, 2015; Stacey, Helme, & Steinle, 2001).

Most research about fractions seems to focus on variants of error analysis or confusions such as Padberg's study, and only a few looks on students' reasoning (Norton & Hackenberg, 2010), in particular fraction arithmetic (Siegler & Lortie-Forues, 2015). Students' reasoning about fractions has been identified as one of the areas that needs more research especially in relation to the development of algebraic reasoning (Norton & Hackenberg, 2010). One of the few is Keijzer and Terwel (2001). They describe their study as an analysis of 'growth in reasoning ability' (Keijzer & Terwel 2001, p.53), from informal reasoning to a more formal reasoning. However, reasoning in their study is not clearly defined, something that is not uncommon in research about reasoning (Lithner, 2008; Sumpter, 2013). Reasoning is instead often reduced to some sort of high-quality thinking based on deductive logic, meaning that we can't talk about different types of reasoning, including incorrect ones, and their foundation. There is also a lack of separation between reasoning and argumentation (Sumpter, 2013).

Therefore, in this paper, we use Lithner's (2008) framework where reasoning is seen as the line of thought adopted to produce assertions and reach conclusions in task solving. It doesn't have to be based on formal logic, and it may even be incorrect (Lithner, 2008). The choice is to see reasoning as a product that appears in the form of a sequence, starting with a task (e.g. exercises, tests, etc.) and ending with an answer. Argumentation is considered to be the substantiation, the part of reasoning that fills the purpose of convincing you or someone else that the reasoning is appropriate. To talk about the content of an argument, we look at the relevant mathematical properties of the components in the reasoning. These components are objects (a fundamental entity, e.g. numbers, variables and functions), transformations (a process to an object where a sequence of these transformations is a procedure, e.g. finding a polynomial maxima) and concepts (a central mathematical idea built on a set of objects, transformations and their properties, e.g. infinity concept).

The division between surface and intrinsic properties aims to capture the relevancy of a property depending on the context. This example, provided by Lithner (2008), illustrates this:

> In deciding if 9/15 or 2/3 is larger, the size of the numbers (9,15,2,3) is a *surface* property that is insufficient to consider while the quotient captures the *intrinsic* property. (Lithner, 2008, p.261)

There are two main types of reasoning: imitative and creative mathematical reasoning (Lithner, 2008). Creative mathematical reasoning (CMR) is reasoning that is novel and plausible and has a mathematical foundation, all which imitative reasoning (IR) does not require. IR is a family of different types of reasoning, but in this paper, we will only separate between CMR and IR. Research looking at university students show that most tasks in tests and textbooks can be solved with IR (Bergqvist, 2006; Lithner, 2004). Similar studies have not yet been done at lower secondary level.

In order to talk about affective factors as part of arguments, such as beliefs, we use the notion Beliefs Indications (BI). BI is defined as 'a theoretical concept and part of a model aiming to describe a specific phenomenon', i.e. the type of arguments given by students when solving school tasks in a lab setting (Sumpter, 2013, p.1116), where beliefs are thought of as "an individual's understandings that shape the ways that the individual conceptualises and engages in mathematical behaviour generating and appearing as thoughts in mind" (Sumpter, 2013, p. 1118). In this sense, beliefs are primarily cognitive structures similar to concept images or misconceptions, and what is expressed by the students in their arguments could be interpreted as indication of beliefs.

Method

Data was collected by video recording task-solving sessions that were fully transcribed. The students' written solutions were also part of the data. Three students, two girls and one boy, from a council school (middle-class suburb to a city) participated in the study. They were in grade 9 which is the last year of compulsory school. Their grades were well above average, but not the highest grade, to ensure that they would have the basic knowledge about fractions. The students were asked to solve three tasks in a lab situation. The tasks were designed to have different levels of difficulties and encompass different aspects of division of fractions. Each task was presented one at the time, and the students could stop whenever they wanted. The first task consisted of five subtasks, with the same question posed: Does the answer get bigger or smaller [than the dividend]? Motivate. Then, solve the task.

1. (a) $\frac{3}{4}\Big/\frac{1}{4}=$ (b) $\frac{15}{3}\Big/2=$ (c) $\frac{4}{5}\Big/\frac{1}{10}=$ (d) $\frac{5}{6}\Big/\frac{13}{7}=$ (e) $\frac{2}{3}\Big/\frac{1}{2}=$

In this task, there is a variation between numerators, denominators and ratios (the answers). The reason why the students were asked to estimate whether the answers were going to be bigger or larger than the dividend was to stimulate logical reasoning based on mathematical properties of division with fractions.

2. Decide the fraction that is half of 4/9. Is the answer bigger or smaller [than 4/9]? Motivate. Then, solve the task.

This task aims to check whether the students understand division of a fraction using 'half of'. Similar tasks can be found in textbook from grade 3.

3. Can you divide two fractions and get the result 5? Explain how you are reasoning.

This task aims to generate data to see if the students could work 'backwards' about division of fractions.

The students were asked to think aloud, a set-up that has been used in previous studies (Jäder, Sidenvall, & Sumpter, 2016; Lithner, 2008; Sumpter, 2013). To structure the data, a four-step reasoning sequence was used (Jäder et al., 2016):

1. A (sub-)task is met, which is denoted as task situation (TS).
2. A strategy choice (SC) is made where 'choice' is seen in a wide sense (choose, recall, construct, discover, guess, etc.).
3. The strategy implementation (SI).
4. A conclusion (C) is obtained.

The characterisation of reasoning types is based on analyses of the explicit arguments for strategy choice and implementation, and the reasoning was characterised as CMR or IR following Lithner's (2008) framework. When studying BI, in this study, we look for explicit metacognitive statements in the transcripts of the task-solving session. BI is data carrying information about the person's beliefs, as defined earlier. Data containing traces of the student's argument were marked. They could be local (for instance, a specific strategy choice) or global (e.g. belief about problem solving). Since the data is not triangulated, e.g. with stimulated recall interviews, here we only see it as indicated beliefs. The BIs were gathered in themes using inductive thematic analysis. The themes were checked against each other and back to the original data since the data within the themes had to 'cohere together meaningfully, while there should be clear and identifiable distinctions between themes' (Braun & Clarke, 2006, p. 91).

Results

In total, the three students generated 33 task situations (TS), most of them ($n = 30$) classified as IR and three classified as CMR. Focusing on IR, two types of foundations were used by the students. The first type of foundation that students used in their arguments was a mathematical foundation although wrong and/or not central for the task. The other type of foundation was indicated beliefs (BI). Here, parts of Ida's and Linn's work illustrate the reasoning and the two different types of arguments. 'I' stands for interviewer.

Ida's Work

Ida is trying to solve the following task: Decide the fraction that is half of 4/9. Is the answer bigger or smaller [than 4/9]? Her work is divided in two parts, and both parts will be presented.

In the first part, Ida tries to decide if the answer to the division is bigger or smaller than the dividend. Her conclusion is reached after just a few seconds, and she gives the following supporting arguments:

Ida: Because when you divide it, then Stefan [the teacher] has always nagged about this thing with cakes... Ah if you have... but how should I explain it, if you have a cake and eh then you should divide into four... and then one can't come so then you should divide into three, then the pieces get bigger.

TS 1: $\frac{4}{9}\Big/2 > \text{or} < \frac{4}{9}$?

SC 1: Reference to teacher's description about division of a cake, comparing magnitudes, here ¼ with 1/3.

SI 1: No further implementation.

C 1: Bigger/Larger.

Ida's reasoning, although referring to mathematical ideas about size of fraction, is not based on mathematical properties central to this task.

In the second part of the solutions process, she tries to solve the division. After some time, when Ida has been quiet, the interviewer asks her about the solution:

I: Do you know the solution/how to proceed?

Ida: No, that is what I'm sitting and eh... thinking of... ninths are not something that we have excessively worked with [laughs nervously].

Shortly after, Ida decides to divide the fraction with 4. She soon realises that she can't divide the denominator (9) with 4 without receiving a reminder and stops without any further explanations:

TS 2 How to solve $\frac{4}{9}/2$?
SC 2 Divide the fraction with 4.
SI 2 Starts, but soon discover 4/9 (i.e. 9/4 gives a reminder)
C 2 Must be wrong, ends with no further arguments.

In this part of the solution, there are no arguments verifying and/or supporting the strategy choice or the implementation, but there is some information about Ida's beliefs about mathematics and mathematics education. She says that 'ninths are not something that we have excessively worked with' as if there is a difference regarding mathematical properties between different types of fractions, and her ability to reason is restricted to what fractions that have been dealt with in mathematics class.

Linn's Work

Linn is trying to solve task 1b: $\frac{15}{3}/2$ =. Does the answer get bigger or smaller [than the dividend]? Motivate. Then, solve the task. Her work is divided in two parts. Here, we focus on the first part since the second part was classified as a CMR.

Linn starts by saying that there are five 3s in 15 and then asks the observer if she can take 'like 5 divided with 2 then?'. She gets no answer from the observer. Linn continues by saying it feels like this should be the case since there are five 3s in 15 and then you should take 5 divided by 2. However, when she starts to implement this strategy, she first starts to write down $\frac{15}{3}$ = 5, i.e. with a horizontal line. When she sees this, Linn says 'Whoopsie, now it got a bit wonky'. She explains that $\frac{15}{3}$ means fractions, and when you write it as 15/3, then it means division. Since she was supposed to divide 15 with 3, she has to write it as a division, because the signs, according to her, have different meanings:

TS 1 How to solve $\frac{15}{3}/2$?
SC 1 Since $\frac{15}{3}$, according to Linn, differs from 15/3, she interprets it now as a division.
SI 1 15 is first divided by 3 and then with 2. Writes it down and performs the division in her head.
C 1 2.5.

The reasoning is based on arguments signalling that her understanding of fractions is dependent on her understanding of division or more specifically the signs used in division. Despite that Linn thinks that the fraction $\frac{15}{3}$ can be performed as a division, she needs to rewrite it to 15/3 before it makes sense. This is interpreted as an indicated belief about difference between division and fraction regarding signs.

Discussion

In this study, three students' solutions were analysed focusing on arguments when the reasoning was classified as IR. Even though there were only a small number of students who participated, which of course means that there are limitations with this study, we can see some results that points to various directions.

First and foremost, the results indicate that the students were not used to formulate arguments based on mathematical properties. The number of reasoning sequences that were classified as CMR was 3 out of 33. Instead, the arguments were anchored in what the students think would be the correct algorithm or a 'feeling' what could be a plausible conclusion, most often with a reference to previous experiences. The mathematical foundation needed for CMR was then replaced with two different themes. The first theme was incorrect mathematical foundation. This is similar to the errors that have been reported in previous research, e.g. Padberg (1989 in Engström, 1997). However, we would like to distinguish what is imitative reasoning and what is incorrect mathematical foundation. When studying students' conceptual knowledge about fraction arithmetic, Siegler and Lortie-Forgues (2015) concluded that although demonstrating ability to perform procedures, students showed weak conceptual understanding of multiplication and division of fractions below 1. This could be interpreted that students can solve tasks with imitative reasoning but at the same time not be able to create mathematical reasoning based on central mathematical properties. Therefore, we look at the arguments. In this paper, we see similar patterns that have previously been described as 'confusion' that arises from when students are trying to make sense of number concepts including fractions (Stacey et al., 2001). And, according to Siegler and Lortie-Forgues (2015), we can add algorithms to this mess: the students' produced imitative reasoning based on incorrect mathematical foundation, but importantly they did not know how to proceed when algorithms did not behave as the students thought they would. Such behaviour has been observed in previous research in mathematical reasoning (Jäder et al., 2016; Sumpter, 2013).

The other theme, which was indicated beliefs (BI), might provide more information about this confusion. The BIs were about mathematics such as there is a difference between division and fractions, although it supposed to be interpretations of the same concept, *a/b* (c.f. Brousseau et al., 2007; Marshall, 1993). This would imply that the students, although in grade 9 and last year of compulsory schooling with relatively high grades, have not learnt *a/b* both as a quantity and as a division, but also as Nunes and Bryant (2009) emphasised the relationships between these two views. If so, this would indeed be a restriction when trying to perform algebraic reasoning (c.f. Norton & Hackenberg, 2010). Imitative reasoning can be considered to be very unproductive even though it could help you to solve a lot of tasks in little time (Lithner, 2008). Bergqvist (2006) explains this further, focusing on algorithmic reasoning which is a part of IR:

> Using algorithmic reasoning is not a sign of lack of understanding, since algorithms are frequently used by professional mathematicians. [...] Algorithmic reasoning is however *possible* to perform without any understanding of the intrinsic mathematics. (p.34)

In this way, an algorithm is designed to avoid to create meaning and understanding (Lithner, 2008), meaning that you could use it without knowing what you are doing.

Another type of beliefs that was indicated was about mathematics education. When Ida says that that 'ninths are not something that we have excessively worked with', it could be interpreted that different fractions have different types of mathematical properties, which would be a belief about mathematics, but also that her reasoning is limited to the fractions that have been in focus in mathematics class. This could be a belief about expectations (Sumpter, 2013): you work with tasks that you have seen in class. In combination, these two indicated beliefs appear to be restricting Ida's possibilities to even start a CMR, even though the task is relatively easy.

Since there are few studies about reasoning about fractions (Norton & Hackenberg, 2010), the implications from this study are two. First, it appears that it is not enough to focus on error analysis in research about students' work in fraction since their arguments also consist of other aspects, here called BIs. Siegler and Lortie-Forgues (2015) stress the importance of working with well-chosen problems including follow-up discussions and not just 'standard' textbook tasks when learning fraction arithmetic. This could, perhaps, not just help students to develop their conceptual understanding but also prevent the creation of unproductive and/or incorrect mathematical beliefs.

Second, when teaching about fractions, it seems it is not enough to focus on fraction as a quantity and as a division. Just as Nunes and Bryant (2009) concluded, the relationships between these two views also need to be in focus since the transfer, as illustrated with Linn's work, is by no means straightforward.

References

Bergqvist, E. (2006). *Mathematics and mathematics education: Two sides of the same coin.* PhD theses. Umeå Universitet, Umeå.

Braun, V., & Clarke, V. (2006). Using thematic analysis in psychology. *Qualitative Research in Psychology, 3*, 77–101. https://doi.org/10.1191/1478088706qp063oa

Brousseau, G., Brousseau, N., & Warfield, V. (2007). Rationals and decimals as required in the school curriculum part 2: From rationals to decimals. *Journal of Mathematical Behavior, 26*, 281–300. https://doi.org/10.1016/j.jmathb.2007.09.001

Engström, A. (1997). *Reflektivt tänkande i matematik. Om elevers konstruktioner av bråk. [Reflective thinking in mathematics. On students' constructions of fractions.] (Studia psychologica et paedagogica, 128).* Stockholm: Almqvist & Wiksell International.

Fennel, F. S., Faulkner, L. R., Ma, L., Schmid, W., Stotsky, S., Wu, H.-H., et al. (2008). *Report of the task group on conceptual knowledge and skills*. Washington DC: US Department of Education, The Mathematics Advisory Panel.

Jäder, J., Sidenvall, J., & Sumpter, L. (2016). Students' mathematical reasoning and beliefs in non-routine task solving. *International Journal of Science and Mathematics Education*. https://doi.org/10.1007/s10763-016-9712-3

Keijzer, R., & Terwel, J. (2001). Audrey's acquisition of fractions: A case study into the learning of formal mathematics. *Educational Studies in Mathematics, 47*(1), 53–73. https://doi.org/10.1023/A:1017971912662

Kieren, T. E. (1976). On the mathematical, cognitive and instructional foundations of rational numbers. In R. Lesh (Ed.), *Number and measurement* (pp. 101–144). Columbus, OH: ERIC/SMEAC.

Kieren, T. E. (1993). Rational and fractional numbers: From quotient fields to recursive understanding. In T. P. Carpenter, E. Fennema, & T. A. Romberg (Eds.), *Rational numbers: An integration of the research* (pp. 49–84). Hillsdale, NJ: Lawrence Erlbaum Associates.

Lithner, J. (2004). Mathematical reasoning in calculus textbook exercises. *Journal of Mathematical Behavior, 23*, 405–427. http://dx.doi.org/10.1016/j.jmathb.2004.09.003.

Lithner, J. (2008). A research framework for creative and imitative reasoning. *Educational Studies in Mathematics, 67*(3), 255–276. https://doi.org/10.1007/s10649-007-9104-2

Marshall, S. P. (1993). Assessment of rational number understanding: A schema-based approach. In T. P. Carpenter, F. Elizabeth, & T. A. Romberg (Eds.), *Rational numbers: An integration of research*. Hillsdale, NJ: Lawrence Erlbaum Associates.

Norton, A., & Hackenberg, A. J. (2010). Continuing research on students' fraction schemes. In L. P. Steffe & J. Olive (Eds.), *Children's fractional knowledge* (pp. 341–352). NY: Springer.

Nunes, T., & Bryant, P. (2009). Key understandings in mathematics learning. paper 3. Understanding rational numbers and intensive quantities. Access 4 Apr 2016 from http://www.nuffieldfoundation.org/fileLibrary/pdf/P3_amended_FB2.pdf

Siegler, R. S., & Lortie-Forgues, H. (2015). Conceptual knowledge of fraction arithmetic. *Journal of Educational Psychology, 107*, 909–918. https://doi.org/10.1037/edu0000025

Stacey, K., Helme, S., & Steinle, V. (2001). Confusions between decimals, fractions and negative numbers: A consequence of the mirror as a conceptuals metaphor in three different ways. In M. van den Heuvel-Panhuizen (Ed.), *Proceedings of the 25th conference of the International Group for the Psychology of Mathematics Education* (Vol. 4, pp. 217–224). Utrecht: PME.

Sumpter, L. (2013). Themes and interplay of beliefs in mathematical reasoning. *International Journal of Science and Mathematics Education, 11*(5), 1115–1135. https://doi.org/10.1007/s10763-012-9392-6

Index

H. Palmér, J. Skott (eds.), *Students' and Teachers' Values, Attitudes, Feelings and Beliefs in Mathematics Classrooms*,
https://doi.org/10.1007/978-3-319-70244-5

MIX
Papier aus verantwortungsvollen Quellen
Paper from responsible sources
FSC® C105338

If you have any concerns about our products,
you can contact us on
ProductSafety@springernature.com

In case Publisher is established outside the EU,
the EU authorized representative is:
Springer Nature Customer Service Center GmbH
Europaplatz 3, 69115 Heidelberg, Germany

Printed by Libri Plureos GmbH
in Hamburg, Germany